About Island Press

Since 1984, the nonprofit organization Island Press has been stimulating, shaping, and communicating ideas that are essential for solving environmental problems worldwide. With more than 1,000 titles in print and some 30 new releases each year, we are the nation's leading publisher on environmental issues. We identify innovative thinkers and emerging trends in the environmental field. We work with world-renowned experts and authors to develop cross-disciplinary solutions to environmental challenges.

Island Press designs and executes educational campaigns, in conjunction with our authors, to communicate their critical messages in print, in person, and online using the latest technologies, innovative programs, and the media. Our goal is to reach targeted audiences—scientists, policy makers, environmental advocates, urban planners, the media, and concerned citizens—with information that can be used to create the framework for long-term ecological health and human well-being.

Island Press gratefully acknowledges major support from The Bobolink Foundation, Caldera Foundation, The Curtis and Edith Munson Foundation, The Forrest C. and Frances H. Lattner Foundation, The JPB Foundation, The Kresge Foundation, The Summit Charitable Foundation, Inc., and many other generous organizations and individuals.

The opinions expressed in this book are those of the author(s) and do not necessarily reflect the views of our supporters.

OVER THE SEAWALL

OVER THE SEAWALL

Tsunamis, Cyclones, Drought, and the Delusion of Controlling Nature

Stephen Robert Miller

ISLANDPRESS | Washington | Covelo

Library of Congress Control Number: 2023934452

All Island Press books are printed on environmentally responsible materials.

Manufactured in the United States of America
10 9 8 7 6 5 4 3 2 1

Keywords: adaptation, aridity, Arizona, Bangladesh, climate change, climate solution, colonialism, Colorado River, desalination, development, disaster, embankments, extreme heat, farmer, foreign aid, geoengineering, Green Revolution, infrastructure, Japan, maladaptation, nature-based solutions, seawalls, technosolution, traditional knowledge, urban growth

For Emily, wherever I may find her

"Our misery is the work of man."

—**GEORGE SWANK**

Survivor of the Johnstown flood, 1889[1]

Contents

Introduction

Lately, I've been thinking about the lost city of Ys. As the story goes, it was an ancient settlement in a bay on the northwest tip of France. Its lord was the pious king Gradlan, who had a sinful degenerate of a daughter named Dahut. Ys had been built on wetland reclaimed from the sea. Its people encircled their city with a dike to protect it from the ocean. Gradlan kept the dike locked tight, except to drain the land at low tide, and wore its key on a chain around his neck. In this way, Ys became a wealthy city with a marble palace, a vibrant arts scene, and plenty of wine. Perhaps too much wine.

Disgusted with the drunken wickedness of its people, God decided to smite Ys. He sent a demon disguised as a red-bearded prince to call on Dahut and convince her to open the gate so he could enter. One night while Gradlan slept, Dahut tiptoed to her father's bedside and slipped the key over his head. Then she ran through the quiet streets to her mysterious suitor. But in her love-drunk state, Dahut made a fatal error: She opened the gate at high tide. In a near instant, the Celtic Sea poured in, and Ys was wiped clean off the map.

French fishermen say they've caught glimpses of a sunken city in the Bay of Douarnenez, and historians reckon Ys probably existed. Some versions of the story place it on what had once been fertile land, perhaps on the now submerged flat that stood above the water during the Last Glacial Maximum, when global sea levels were lower and Europe extended farther into the Atlantic. If so, Ys had probably battled the encroaching ocean for years, generations even, as the planet's ice caps melted and its beaches receded.

What gets me is that these ancient Bretons had such an advanced system for coastal water management. They must have been at work on it for some time, because you don't add a lock until you've tried a door without one, and you don't cut a door before first building a wall. Perhaps their ultimate demise was simply a result of living in the wrong place at the wrong time, or maybe it was blasphemy, in the form of believing they could control the sea. We should take note.

Not too long ago, adaptation to climate change was something of a fringe idea, resisted by people who worried it would distract from the more pressing need to the mitigate greenhouse gas emissions that cause global warming. Now that ship has sailed. Environmental catastrophes that we had never thought possible—or that some of us weren't used to seeing so close to home—have become common. Some combination of erratic weather, late-season wildfire, catastrophic tropical storms, blistering drought, rising seas, and ecosystem collapse affects every inch of the globe. And knowing this, people the world over are pushing for what had once seemed like surrender: They are adapting.

On farms and along coasts, in busy metropolises and quiet villages, they are digging canals, erecting seawalls, dredging rivers, wiring microgrids, desalinating water, planting trees, piling embankments, banking seeds, manipulating DNA, and seeding clouds. Our species has adapted to its environments forever, but never have so many of us undertaken such a widespread, coordinated, and hurried attempt to remake our world. We're living through the first wave of climate adaptation; there will be mistakes.

This is a book about unintended consequences, about fixes that do more harm than good and the folly of overconfidence. Academics call it maladaptation; in simple terms, it's about solutions that backfire.

In 2009, Jon Barnett and Saffron O'Neill, geographers at the University of Melbourne, published the definitive explanation of the phenomenon. They also laid out some criteria for its identification.[1] Adaptation to climate change goes bad, they reasoned, when it increases greenhouse gas emissions, saddles vulnerable people with a disproportionate burden, forces us to give up on other options, reduces our incentive to adapt, or sets us on a path with fewer choices. From what I've seen, a calling card of maladaptation is that it makes people feel safe in harm's way.

Perhaps it shouldn't come as a surprise that we blunder around with adaptation. After all, we're doing it backwards. Consider the horned lizard. It didn't foresee a tough life in the desert and opt for a thick skin to keep from drying out; it adapted the trait over eons *in response* to its environment. Adaptation makes sense only in retrospect. We're striking out into new territory by trying to anticipate what's coming, and we can't be absolutely sure of what the future holds. When we rely on faulty or partial predictions, we risk not only making the wrong choice but closing ourselves off from better options. As the years pass, the effects of our decisions become so entangled with the changes to our environment that we can't tell whether we're adapting to climate change or to our own past choices.

Americans out West already are familiar with this. Each year, we breathe in the result of the US Forest Service's decades-long commitment to the 10 am policy.[2] This Depression-era order sought to stamp out any wildfire, no matter how large or far from where people were living, by 10 am the morning after it was discovered. In the years before the policy was put in place, a series of devastating blazes had consumed millions of acres across Western states. In their smoldering wake, they left a mandate to do *something*, and fast. Since then, the policy and the antagonistic sentiment toward fire that it embodied have resulted in huge and unbroken tracts of forest with an unnatural abundance of fuel. An effort to protect the economic value of Western forests inadvertently created infernos that are larger and harder to fight.

When I set out to write this book, New York City was still reeling from the effects of Hurricane Sandy in October 2012. They were debating a particularly audacious method for protecting themselves from sea-level rise. Among other options, the Army Corps of Engineers had proposed a 6-mile-long wall spanning Hudson Bay from Queens to New Jersey.[3] Critics warned that its construction could lead to a slew of new problems, especially in low-income neighborhoods where people already suffered from storm flooding. In the bay, the wall risked altering or blocking the flow of Hudson River sediment and impeding migratory marine life. On land, models showed that when the wall's massive gates were shut, wastewater would back up into the city. So the decision facing many New Yorkers was whether they were willing to stew in their own filth behind a barrier that, because of the rate of sea-level rise, would probably be inadequate by the time it was finished.

New York risked shelling out an estimated $119 billion on a solution that could easily make things worse. And in that, it was not alone.

Similarly treacherous coastal protections have been proposed in Miami, Florida, and Galveston, Texas, and seawalls already cause problems in Orange County, California.[4,5] Across the country's parched interior, there are calls for desalination plants that generate huge amounts of greenhouse gases. In Colorado, there are experiments with cloud seeding, which hope to bring rain in one place but risk causing floods or drought in others.[6] There are also insurance schemes that invisibly soften the economic losses of hazards, leaving people less concerned with the dangers of living at the edge of a tinderbox forest or on the sinking coast. And there are subsidies that encourage farmers to stick with crops that can't handle the heat.

If implemented properly, any of these reactions might be tools in a successful adaptive strategy, but they often aren't done well. Their plans are shortsighted and ill-conceived. They get shortchanged by governments with fleeting priorities and co-opted by businesses focused on making a buck. Even under the best intentions, they shove on like bulldozers with blinders, detached from the nuances of the people and places they're supposedly helping. Under those circumstances, adaptations don't lift the burden; they shift it onto people of another class, place, or time. Often enough, though, the shortcoming isn't so nefarious. We all know how hard it can be to break a habit; once a system is in place, the easiest thing to do is to keep using it. As I saw time and again while researching this book, bad habits result in technological lock-in, where new hardware and software are built to plug into existing infrastructure. The same is true of our mindset: Our current perspectives and expectations determine our future decisions.

I got a whiff of this happening long ago in the Arizona desert where I grew up, but I didn't have a word for it. Millions of people like my family were moving to an environment with obvious limitations. They were lured by development boosters and industries that promised safety and wealth even as headlines reported critical resources running dangerously short. Something was giving us all the confidence to settle someplace where access to life's most basic need was a legitimate concern. I chalked it up to good old American grit, tenacity, and determination. Then, years later, I came across the story of a coastal Japanese city where people had stood atop a seawall to watch the approach of the tsunami that killed them. They believed—wholeheartedly—that the huge concrete barrier would save them.

Here it was in a nutshell: people harmed by the very thing built to protect them. Although their story was not directly connected to climate change, I recognized it as a cautionary tale for that arena. As compared to the creeping onset of sea-level rise or drought, the immediacy of a tsunami helped me wrap my head around this wonky concept. Then, when I made my first visit to Bangladesh in 2019, I stared down its most impressive incarnation and uncovered the often unseen influences of race, religion, and money.

~

The three stories that follow come from disparate locales, widely varying cultures, and wholly unique circumstances, but each explores an adaptation gone awry. In trying to understand how that happened, I discovered that they have plenty in common. To start with, the people behind them were myopic in their predictions and overconfident in their designs. Some were downright corrupt, others just woefully ignorant. They often confused their goals, conflating profit with success, and in each case they ignored warnings.

Many small-scale adaptive efforts and government policies flop; we must be willing to fail on some scale to find what works. I've chosen to focus on three glitzy techno-infrastructural solutions, because they cost a fortune and require bureaucratic mobilization on a warlike scale. As a result, they typically form the centerpiece of political legacies and so fall prey to ambition. Large-scale infrastructure also tends to erase the past, which we need in order to tell where our progress is trending. And it confines our future by imposing its will on our choices, like bumpers in a bowling lane.

The stories in this book trace the histories of engineering marvels that were once deemed too smart and too big to fail. If that were true, they wouldn't be here.

Concrete and steel, copper and rebar, computer models and simulations—there are plenty of those things in the following pages, but what has interested me most are the people. Behind every wall or canal are rich cultures with fraught histories and complex allegiances. In trying to understand the nuances of local economics, colonialism, religion, and societies of which I had been (and still am) embarrassingly ignorant, I found stories of people trying to solve problems. That's why I've been thinking about the lost city of Ys.

Adaptation is not a solution; it's a practice over time. Although we may find ourselves closer to the mark, we will never arrive, because something is always in flux. Some of the decisions I discuss in this book were made outside the context of climate change, and so it's unlikely that the people who made them thought of what they were doing as adaptation. More likely, they were only fixing a problem. But isn't that all we've ever done? Not to downplay the flagrant greed of corporate executives or the abject failure of politicians who have known for too long that burning huge quantities of fossil fuels damages the lives of many while catastrophically altering the planet for all, but the fact is that no one set out to cause global warming. It's been a consequence of each generation seeking solutions to everyday challenges: how to move faster, how to grow more food, how to do more with less. Each has tried to improve its lot, and the compounding results have led us here. If anything, I hope this book will rattle our thinking to break that habit.

In that vein, you'll find that this book shies from championing solutions. It seems antithetical for someone criticizing simple fixes to complex problems to offer more of the same. My aim is to elucidate the problem and to show how it has played out in varying contexts so that we might recognize it at home, before we commit to a path we later regret. Because despite all we know about how climate change will affect our world, there's still plenty we can't yet imagine. The limits of our understanding mean that tradeoffs are inevitable. There are no futures without regret. In a climate of such intense uncertainty, we wager the future against the present.

I used to write sentences like that last one from a comfortable distance. And then, partway through working on this book, I found out that I was going to be a father. The news has challenged my cynical bent and has caused me to think about the risks I'm willing to take, because they aren't just mine anymore. As cribs and car seats and diaper pails have materialized in my home, I've also been thinking about where I'd like to raise my family. I live in Colorado now but have fond memories of the many years I spent in Tucson, Arizona. The landscape is astonishing, the food is great, housing is affordable, and taxes are low. Given all I know about the threats bearing down on the desert, though, I hesitate to move home, let alone establish my family there. In fact, I doubt I'd even consider it if not for a slender line of concrete that carries precious river water across hundreds of miles to the center of the drought-ridden desert. That's the effect these bad adaptations can have: They turn suspect options into safe bets.

In the end, it comes down to choice, which is still the most powerful adaptation available to any of us. For an expecting parent, this choice about where to settle presents a new moral hazard. Do I heed the warnings? Or, like those at Ys, do I accept the risk, build a wall, and hope that it is never opened?

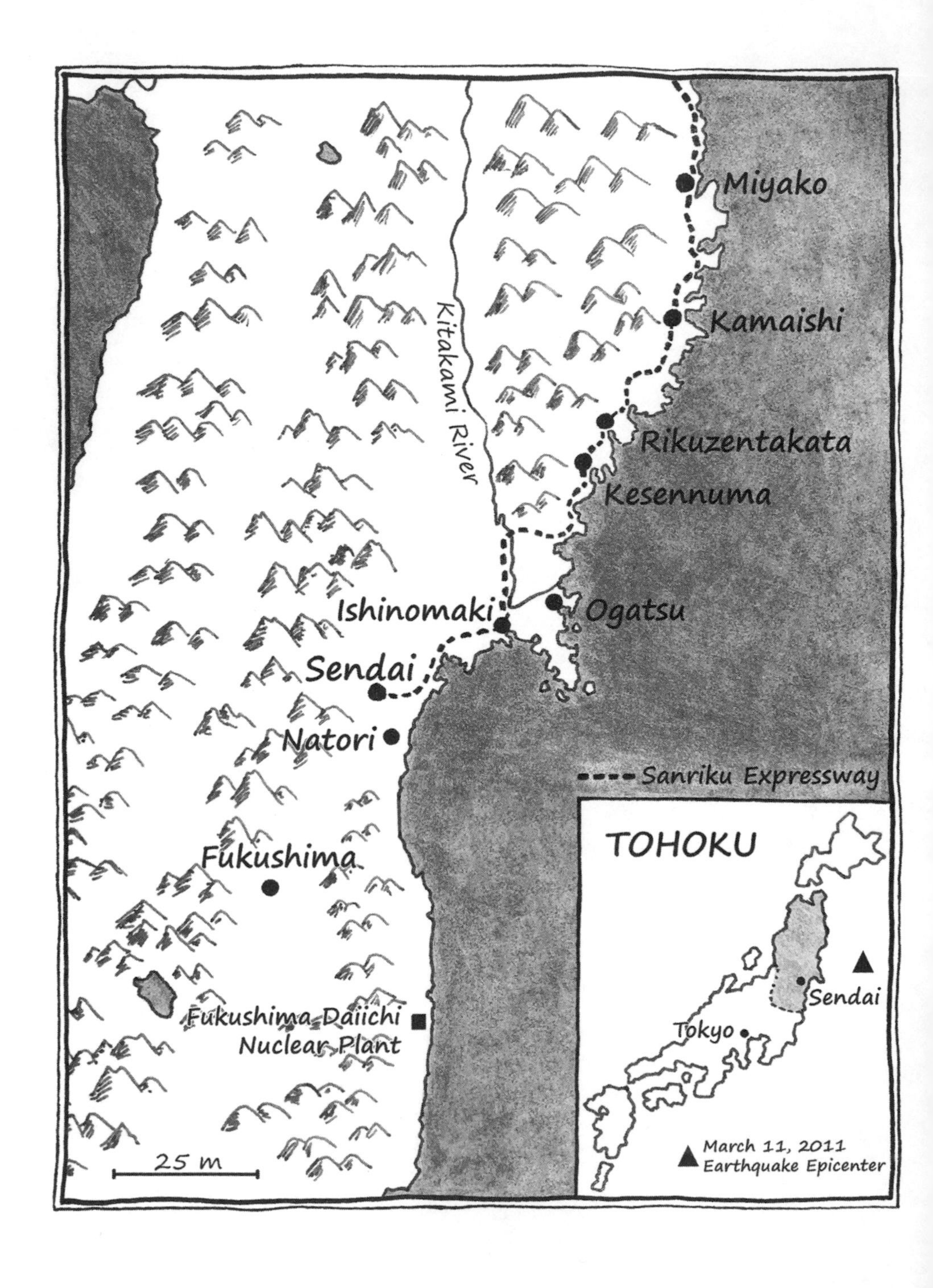
Miyako
Kamaishi
Kitakami River
Rikuzentakata
Kesennuma
Ishinomaki
Ogatsu
Sendai
Natori
Sanriku Expressway
TOHOKU
Fukushima
Sendai
Tokyo
Fukushima Daiichi
Nuclear Plant
25 m
March 11, 2011
Earthquake Epicenter

Part I

Soutei-Gai

Northeastern Japan

After a tsunami flooded the small fishing community at Shibitachi Bay in northeastern Japan on March 11, 2011, government engineers proposed to build a 32-foot-high seawall along the waterfront. Locals worried that the proposed infrastructure, which would cover the generations-old port in concrete, would cut fishermen off from the sea and damage the community's way of life. They begged officials to keep the new wall no higher than 16 feet. In July 2022, my guides, Takaharu Saito and Ulala Tanaka, walk past the construction of the seawall, which will be 28 feet high.

"The world of science is not a world of reality,
it is an abstract world of force."
—Sir Rabindranath Tagore[1]

The first of the wooden Buddhas fell from its shelf at quarter to three that afternoon. Then a heavy bronze candlestick rolled off the altar and hit the floor with a thud. A porcelain vase toppled, shattering into pieces and shedding flower petals across the new tatami mats. Gold-leafed curtains and bright prayer flags jerked violently from the ceiling. Faded paintings slapped their walls. Paper doors rattled their wooden frames. In his office down the hall, a clean-shaven and bald priest grabbed the edges of his desk. He held tightly as the shaking went on and on and on. Paper lamps swayed like lanterns on a rough sea. A hammered bronze bowl tumbled from its perch. Outside the ancient temple, bells rang erratically. Clay tiles shook loose and slid down the roof's steep vault until they shattered on the lawn. Below the steep hill where the temple perched, screams floated up from Ishinomaki.

On March 11, 2011, the ground shook for six minutes, far longer and more powerfully than anything the priest had ever experienced. It bucked fiercely, angrily, like some restless dog shaking dirt from its fur. Six minutes was long to understand what was happening, long enough to brew a pot of coffee and wonder whether the world would ever be still again.

When at last the shaking settled, the priest bolted from his study toward the temple's main hall, where he found the once solemn retreat in shambles. Its ornaments and artifacts were strewn about as if a bull had been let loose. He stepped carefully around the wreckage, feeling shards of pottery crack under his weight, and headed for the Dogenin Temple's most sacred possession, a gilded statue of the Buddha that had been gifted many years ago. Amazingly, it survived, and the priest felt a moment of relief. But only a moment, because anyone who lived on the coast of Japan knew that an earthquake was just the beginning.

Outside the temple, the world was coming undone. Forty-five miles off Japan's eastern coast and 15 miles under the surface of the sea, there is a trench where the Pacific plate grinds under the Eurasian plate carrying the Japanese islands. Here, centuries of geologic tension had just been released.[2] In that sudden moment of relaxation, Earth's crust let loose enough energy to power Los Angeles for a year. The planet's axis shifted, shortening our days by 1.8 microseconds.[3] Japan's main island, Honshu, jolted 8 feet east, while a 250-mile stretch of its coastline dropped 2 feet. Onshore, land slid and soil liquefied. Offshore, a 110-mile-wide section of seabed shot upward 26 feet, sending a tsunami barreling toward the coast.

Despite what the vibrant, traditional woodcuts would have us believe, a tsunami is not one enormous, foaming wave but a succession of swells that build upon each other. The first of these took thirty to forty minutes to reach land, depending on location. This agonizing wait challenged Japan's state-of-the-art alert system, tested the patience of millions of coastal residents, and sealed the fates of thousands.

The quake's epicenter was just northeast of Sendai, the largest city in Tohoku. This oft-overlooked region of northeastern Japan stretches from gently sloping plains in the south to the sawtooth coves and cliff-walled bays of northern Sanriku. Waves bore down on the flat lower coastline in miles-long, white-capped curtains of death. They spilled over wide beaches, sweeping up thousands of wind-twisted black pines whose roots has been loosened during the shaking and drove them like battering rams 6 miles inland. Along the way, this barrage of debris caught up with terrified people in flight. In cars and on foot, they were gathered into a thick gray sludge that tore through wooden barns, greenhouses, roads, airports, schools, and neighborhoods, blending the human with the inhuman until there was no telling them apart. Waves breached the fortifications at the Fukushima Daiichi Nuclear Power Plant, causing a reactor meltdown that sent radioactive material floating on the wind and the Pacific Ocean. The narrow inlets of northern towns funneled the water, concentrating its impact as if through a hose on the limited flat land between mountains and sea. Here, waves ran up to heights of 130 feet and took their greatest toll as residents drowned in desperate flight up slippery hillsides.[4] As thousands fled Ishinomaki, their cars backed up in total gridlock. Some jumped out and attempted to run. Others remained in their seats as their vehicles were swept up and committed to the churning dark mass that pummeled forward.

The Dogenin Temple had stood on its hill for 950 years. Hundreds of parishioners prayed there regularly. The priest, Hidemichi Onosaki, and his wife, Miki, held funerals, festivals, and potluck dinners. At about half past 3 pm, with the power out and water faucets dry, Miki was on her knees in the main hall, collecting sand spilled from an incense holder, when the first survivors arrived.

"Many had been swallowed by the tsunami and were dazed. Some were carried in on stretchers," she recalls. "There were a number of people who were soaking wet. I gave them all my clothes—all the ones that fit." Then she walked outside to see what they were fleeing from.

This far north, the island was still locked in winter's grip, and she looked toward the mayhem unfolding below through the bare limbs of an oak tree. "You could see it right in front of your eyes. The tsunami did not look like anything in this world. It was like watching an action movie. The sound was a great rattling sound, crunching and crunching. It sounded like the whole town was breaking down." Fat snowflakes began to fall. They glinted in the light of fires roaring on floating piles of debris. Aftershocks tormented the city even as waves returned mercilessly. All through the late afternoon and early evening, people climbed the slippery drive to Dogenin, shivering and traumatized. As the sun set, Miki hurried too cook for them in the fading light.

At least one hundred people fled up the hill to Dogenin Temple that first night. The priest recognized many, but most were strangers to him. More bedraggled survivors came the next day, and the next, until by March 14, some 400 people had taken shelter there. They slept side by side on tatami mats in every room but the bathroom. They ate rice and miso soup cooked over propane stoves, and as the fuel dwindled they descended into the obliterated town to pick through the remnants of their own lives for firewood.

Hidemichi Onosaki never turned anyone away, but as the days passed, he knew they needed some method of maintaining order amid the grief. He led morning prayers and group calisthenics. Everyone was given a task, a responsibility to keep them engaged, even if all they could offer was thanks. And he laid out a set of rules. Mostly, these required that everyone treat each other kindly and help out as they could, but his most important rule was the seventh: Always remain grateful.

"Grateful?" one survivor asked Miki. "Listen to me, ma'am. I've lost my wife. I've lost my house; I have absolutely nothing left now. All that's left for me is to die."

Grief in the moment was eclipsed only by the horror and bewilderment about what had happened, and from the years she had spent leading grieving families through funeral rites, Miki knew what a salve work could be. With all those needing help and all that had to be done, she thought, "We have no place for a debris of a man." And she assigned him the task of serving the morning tea.

In the years after 2011, people who sheltered here recalled it with mixed emotions. It was the worst time anyone could remember, and yet the temple, many of the children said, was bright and felt like home. This brings Miki to tears. Eventually, the people left, moving into new homes they built themselves or into the government's temporary housing. But many returned to Dogenin for months afterward. They said the sense of purpose got them through. They said they were helpless during the disaster, but they could control what happened after.

~

By the time I arrive at Ishinomaki, eleven years have passed since one of the most horrific catastrophes in modern history. Like millions of others around the globe, I had watched and rewatched hundreds of witness videos, spending hours descending down a rabbit hole of some foreign grief. I had read the news reports and books, listened to recordings of survivors' stories, mined the academic literature, and spoken with dozens of experts. There was nothing that could have prepared me for what I hear and see here.

Now, I sit cross-legged amid the earthy scent of tatami mats in the Dogenin Temple, drinking iced yellow tea and listening as the soft-spoken Miki tells of how the last of the survivors left in August 2011. It's a clear, hot day in what has been an oppressively muggy summer, and I'm happy to be sitting out of the sun while her Japanese washes over me. During my visit, which occurs while the country remains largely locked down in response to the COVID-19 pandemic, I will be aided by a small army of interpreters who not only translate the words through dozens of interviews but also fill me in on the backstories and cultural nuances that can be so hard to glean secondhand. Today, I'm traveling with Sébastien Penmellen Boret, a French anthropologist who studies death, grief, and burial rituals. While I take in the serenity of the temple's inner *hondo*, he talks with Miki and the priest about the hundreds of Buddhist ceremonies they've held for the 19,747 confirmed victims and 2,500 others who are still recorded as missing after an eleven-year absence.

My eyes wander along the gilded altar, the intricately carved statues and the hanging banners. If not for the photos of children sleeping on this very floor, bundled in dirty winter coats, I couldn't tell anything out of the ordinary had ever happened. Through sliding paper doors, past a garden, and under a tall wooden shrine gate, Ishinomaki spreads beside a glistening bay. Fishing boats slide into port and sailboats cut white lines through the silver water. This is the same vantage from which Miki gazed upon the mayhem unfolding eleven years and five months before. Where she saw the water rise between the homes, lift them up, and crush them together like dollhouses. Where, as night fell, she watched fires flare on islands of floating debris while she received refugees.

Ishinomaki lost 3,173 souls that day, more than any other city in Japan, but not because it was uncommonly vulnerable.[5] The port was well developed, with breakwaters out in the bay to limit erosion and slow waves before they reached shore. A concrete seawall skirted the coast to resist surges and king tides. Its flat face was pockmarked from decades of abrasive tides and rose straight up to about 20 feet in some places. It was no match for the wave that came in 2011, but nothing was. No country was better prepared for a tsunami than Japan, and yet its leaders later admitted to being caught by surprise. Tens of thousands perished. I'm still wrapping my head around it. I'm not the only one.

The disaster lured swarms of journalists, civil engineers, anthropologists, and even tourists, who found stories of immense sorrow and unexpected hope. Fishermen who escaped by sailing into the wave and daughters who were lost while searching for their parents. Forests wiped away and towns rallying to rebuild. A journalist who covers climate change and adaptation, I found myself caught up in the story of a people harmed by the very efforts made to protect them.

Climate change is often discussed as a monolithic crush bearing down on all of us at once, but what we experience is a series of familiar hazards—though more intense, more frequent, and certainly more erratic than we're used to. A tsunami might not be a climate-driven phenomenon, but it carries the hallmarks of one: the breadth and severity of harm. If Japan could fall victim to adaptations built to protect them from a hazard it has known for eons, so could we all.

Early in trying to understand how it had happened, I spoke with a Japanese anthropologist named Shuhei Kimura, who was asking many of the same questions. He told me of a fifty-two-year-old man whom he called Konno-san,[6] who was not home when the tsunami struck but whose wife was. Of the 2,500 residents in their coastal village, she was one of only thirty who perished. Between bouts of unhinged grief, Konno worried that her death had been "embarrassing." *Shameful* might be a better word. The Japanese have a rich vocabulary to describe shades of guilt, and, as I will learn, it's impossible to understand the full complexity of a survivor's feelings. An elder in the community and the head of a respected household, Konno felt ashamed that he had not saved his wife. He worried about how it must have looked to his neighbors. He also wondered why his wife had not led the way to safety, because he expected his family to set an example. Why had she remained while all around her alarms whirred and neighbors fled in panic? He blamed the seawall; the numbers back him up.

In the months after the disaster, a team of anthropologists from Oxford University discovered that the death toll had been highest in communities where the government had recently invested heavily in coastal protections, such as seawalls and levees, and where people had little experience with such waves.[7] They reasoned that, in many cases, Tohoku's seawalls had been maladaptive. Investment in protective infrastructure had not only stimulated settlement in areas that were known to be vulnerable to flooding but also imparted a false sense of security that delayed victims like Konno's wife from fleeing.

Subsequent research showed that despite Japan's long relationship with tsunamis, education campaigns, and regular evacuation drills, coastal residents were as much as 60 percent more likely to stay put when they lived behind a seawall that rose above height estimates for an incoming wave.[8] On March 11, 2011, initial alerts predicted a 9.8-foot tsunami, and many towns were already protected by 13-foot seawalls. In Ishinomaki, the wave reached 22 feet.

Miles north, residents of the Taro district of Miyako City suffered the worst of this bad adaptation. They had constructed a concrete barrier so robust locals dubbed it the Great Wall of Japan.[9] "It was the pride and symbol of Taro. I had never imagined that this seawall could be breached," said one survivor.[10] X-shaped to offer double protection, the Taro wall reached 34 feet high and stretched along 1.5 miles of coastline. It was the largest of its time, and in 2011, people were so confident in its strength that they stood atop it recording images of the wave until the very moment they were washed away forever.

The unique design turned out to be a terrible mistake. The 57-foot wave's energy focused on the center of the X, where it shot up and over the wall onto the town that had settled in its shadow. At Taro, residents had put their faith in technology developed by one of the most advanced societies on Earth. One-hundred-sixty-one victims discovered too late that their trust had been misplaced. So it struck me that in the wake of such catastrophic failure, the city responded by building a new wall, 14 feet taller than the last and yet still more than 8 feet shorter than the 2011 wave.[11]

Seawalls undoubtedly saved untold lives and property in 2011, and since then, Japan has doubled down. Knowing that another tsunami is inevitable, the country has raised roads, elevated and relocated entire towns, planted coastal forests, and rethought its evacuation routes. Some of these were novel measures, but the centerpiece of reconstruction is an old idea: encasing the coast in concrete fortifications. Today, construction has nearly finished on some 400 miles of breakwaters, river levees, and seawalls as tall as 50 feet, at a cost of about $255 billion.[12] As the walls go up and the memory of Tohoku's devastation fades further into the past, people rebuild their homes and settle on raised land in areas that were obliterated by the wave.

That the new walls are so much mightier is encouraging, but I also discovered a growing sense of disillusion toward them. It manifests in heated community planning hearings, protests over the flippant use of concrete on natural coastlines, and the embarrassment that a 52-year-old man feels about his wife's final moments.

I came to Tohoku to learn what had happened and to find what, if anything, the past decade could teach the rest of us about how—or how not—to adapt to this increasingly unforgiving planet. Tsunamis may not be caused by climate change, but, like climate disaster, they are collections of hazards capable of consuming huge areas and countless lives. Each can be explained with math and physics. Mostly, though, they both confront us with the enormous challenge of understanding ourselves. Sitting in Dogenin, I can picture the hundreds of people dragging themselves up the steep approach in the thick, wet snow of a distant March. When all the engineering marvels, techno-wizardry, and gadgets had failed them, victims of a horrific tragedy came seeking the only salvation they could think of: They went to a centuries-old temple upon a hill. Given the threats bearing down on all of us across the world today, the uncountable risks and inconceivable uncertainties, and the comparative frailty of our grandest inventions, I'm left thinking that sooner or later we're all going to find religion.

Life and Death in a Name

In Japan, 125 million people live on 145,937 square miles.[13] The State of California today has a total area of 163,696 square miles and a population of just under 40 million. Unlike in California, of which some 43 percent is used for agriculture, a full 80 percent of Japan's total landmass is covered in steep, forested volcanic mountains. This leaves a paltry supply of flat land to be shared between industries, houses, and farms. Sendai, the Tohoku region's largest city, squats on an agricultural plain about

30 miles south of Ishinomaki, between the mountains and the mouth of a large, crescent-shaped bay. It was settled long ago and once ruled by a powerful samurai whose seventeenth-century castle still haunts a nearby hill. Today Sendai is an industrialized metropolis, home to one of the country's largest fishing ports and most advanced tsunami laboratories, the International Research Institute of Disaster Science (IRIDeS) at Tohoku University.

One of the first things I learn is that the existing walls and breakers had not failed to protect the northeastern coast from the tsunami; they had never been expected to withstand such a wave at all. Nothing had been built to withstand a 130-foot tsunami. Masataka Shimizu, then president of Tokyo Electric Power Company (TEPCO), which owns the Fukushima Daiichi Nuclear Power Plant, said afterward that there was no way they could have predicted even a 50-foot wave. "I hope that a thorough examination of the effects and cause of the unprecedented tsunami will be done in the future," he said. Then, he and many other talking heads went on Japanese media repeating a single pernicious phrase: *soutei-gai*.[14]

The phrase means "beyond expectation" or "beyond all conceivable hypothetical possibilities," which was an apt term for shirking responsibility. Surely, something that existed outside the realm of possibility couldn't have been predicted and therefore couldn't be prevented. This was the phrase TEPCO used to explain the meltdown at the nuclear plant it built beside the sea. Later, citizens of Fukushima discovered that while preparing for future risks, the company had purposefully collected data only on "reasonably expectable and manageable" events while omitting those that seemed unlikely or unmanageable, to avoid the costs of addressing them. Also inherent in *soutei-gai* is a notion held by the country's political and scientific leaders that nothing could destroy their designs. It wasn't just that they couldn't imagine such a large wave; it's that they couldn't imagine anything stronger than what they'd built.

Soutei-gai immediately caught my ear, because it reminded me of another term I'd been hearing so much lately: *the no-analog future.* Reporters and academics use this phrase when referring to the uncertainties surrounding climate change. It implies that we are entering a world unlike the one we're coming from, but it's not so simple. Our past records may look woefully out of date amid unprecedented heat waves and storms, but we have a long history with our planet's wrath and plenty of analogs for adaptation along the way.

I visit IRIDeS because the school is located in the disaster zone and because it's home to researchers who study everything from engineering and ocean bathymetry to folklore and memorialization of the dead. Yuichi Ebina is a stocky and exuberant associate professor who wears black-rimmed glasses and talks with his hands spread wide apart. He specializes in Japan's historic record of disasters, and he recently made a discovery that undermines one of 2011's greatest myths.

In Japan, before 2011, the largest wave in a generation had come in 1960 as a result of the strongest earthquake ever recorded, a 9.5-magnitude shock off the coast of Chile. It took twenty-four hours to cross the Pacific and crested just under 20 feet in some parts. The Chilean Earthquake Tsunami claimed 140 lives and became the standard to which the Japanese government built many of the protections that were in place fifty-one years later.[15] However, as coastal towns undertook the arduous process of cleaning up after the more recent disaster, they discovered clues that exposed the limitations of their expectations.

I trade my shoes for slippers at the door of Ebina's laboratory in Sendai and cross into an open, white-walled room lit by fluorescent bulbs and packed with cardboard boxes. Ebina collects and restores old documents. Sometimes, these are official records from government buildings, but more often lately they come from regular people who find receipts, letters, journals, logbooks, and accounts of all kinds while moving or refurbishing their homes. He receives truckloads of crusty, wilted parchments—wedding notices like dried leaves, ship logs streaked with black mold—often bound with faded ribbon. Writing craft is a big deal in Japan, and family hoarding is a boon for Ebina, who uses these seemingly benign treasures to peer into what he calls "real history," the lives behind the official records.

Japan's Edo period spanned the uncharacteristically peaceful years between 1603 and 1867. Finally united under a samurai-led shogunate after near-constant internal warfare, the country closed its borders to the outside world. Granted isolation and quiet, scholars had time to write, and peasants became literate. The result is a vast compendium of accounts from nearly all aspects of life. As there has never been a shortage of disaster in Japan, records from this time are rife with accounts of misfortune, floods, storms, and famine. After 2011, Ebina focused on one particular event: a tsunami that had occurred exactly four hundred years earlier.[16]

Historians call this wave Keicho Sanriku: *keicho* for the time period and *sanriku* for the stretch of far northern coastline where its impacts were thought to have been contained. Ebina was searching the weathered pages of a centuries-old copy of a diary that had been written in 1612 when he came across one of the earliest known examples of the word *tsunami*. Written in vibrant calligraphy on fine parchment, or *washi*, it references Tohoku's storied *daimyo*, or feudal lord, Date Masamune and tells the story of two samurai who survived a freak wave:

> Masamune wants fish. Two samurai receive the order. They round up fishermen. The fishermen balk because the sea has a strange color and the skies look ominous. One of the samurai insists on obeying the daimyo's order. All set out in a boat. Soon it meets the tsunami, which drives it inland into the crown of a pine tree. The waves also sweep away entire villages along the shore. Later, after the water recedes, the men clamber down from the tree. Scanning the shore, they realize that they too would have been swept away had they not gone fishing for Masamune.[17]

Ebina was confused. The diary had been written by a shogunate aide traveling around Sendai, far south of where the Sanriku wave was thought to have hit. He checked its account against log entries from an armada captained by the Spaniard Sebastián Vizcaíno, who, several years after naming San Diego Bay, happened to be sailing past Japan. Vizcaíno's account also reported a great wave off Sendai. Then, in 2013, a reconstruction crew digging through debris on the Sendai plain exposed a cross-section of earth that clearly showed a layer of sand deposited between similar bands known to have been laid down by two other massive tsunamis: the Jogan tsunami of 869 and the Great East Japan tsunami of 2011.

Taken together, this evidence points to the Keicho Sanriku tsunami being an order of magnitude larger and more destructive than previously imagined. Now, Ebina wants to change its name to *Keicho Oshu—Oshu* being an old term used for the entire Tohoku region—to acknowledge how much of the coast had been affected. When it comes time to build, rebuild, and protect ourselves from what threats lurk on the horizon, he says, "We cannot just depend on scientific data." He passes me a bundle of decayed, hand-brushed tax ledgers. "We need to respect these old records. Disaster science needs to value history. A modern society becomes vulnerable when it forgets its past."

Based on records of the Jogan Earthquake in 869 and the Meiji Sanriku tsunami of 1896, which killed some 22,000 people, seismologists warned in 2001 that Tohoku was probably due for a once-in-a-thousand-years event. But there's something debilitating about a millennium-wide timeframe, and no one had acted on the information. Ebina believes that if they had known the interval between such disasters was a mere 400 years, people might have thought differently about building in the footprint of past disasters. As it happened, in the decades preceding 2011, they constructed homes, oil refineries, nuclear plants, and elementary schools atop the wreckage of the 1611 event, only to realize the error when they had to dig out again.

"At the moment, when there's a project of rebuilding a community, there's not necessarily references made to long-term plans, how people adapted to these risks in the past," Ebina tells me. This represents a remarkable blind spot for a country with so much history and so much to lose by ignoring it. The omission not only influenced how coastal cities were built before 2011 but colored decisions made in the immediate aftermath. Doubting the likelihood of such an event, the government had no plan in place should one arise, and so, when the time came to rebuild, it fell back on old ideas, even if its people were ready for something new.

Mental Ninjutsu

Nothing that was revealed on March 12, 2011, lent itself to careful reading of decayed family documents. The waves had wrought devastation along 1,242 miles of coastline, an extent just shy of the US seaboard from Tijuana to Neah Bay. Half of those who perished were in Miyagi Prefecture, which stretches from Fukushima to Kesennuma, including Sendai and Ishinomaki, and across from the central Echigo Mountains to the Pacific Ocean. This, Tohoku's most populous prefecture, lost 20 percent of its population.

Afterward, 320,000 survivors needed somewhere to stay.[18] Power, gas, and water lines were cut off, and telephone coverage was limited. Five hospitals and 1,014 schools had been destroyed. More than 300 culturally significant properties had been lost. One hundred thirty-six kids had become orphans. Nine hundred twenty-one had lost a parent. There were 3,312 nonfunctioning traffic lights and 10 million tons of debris, which is fourteen times more than what Miyagi generates in a year. More than one tenth of the farmland had been inundated, and all 142 fishing ports had been seriously damaged. Refrigerated storehouses had split open, spilling tons of frozen fish that thawed and rotted in the open for weeks. Aftershocks riddled Miyagi for months, while the surviving members of governments clawed the coast back to life and scientists scrambled to prepare for the next wave.

Japan has been on the forefront of seismic and tsunami prediction science since the 1970s, but by all accounts 2011 caught it by surprise. Anawat Suppasri is a coastal engineer and tsunami expert at IRIDeS who has studied how the country's coastal defenses reacted, and he specializes in forecasting the next event. He explains complex information with ease, but given his personal history, I worry about sitting in the same room as him. Suppasri lived through the catastrophic 2004 tsunami in Thailand, then moved to Sendai just in time for another.

When an earthquake strikes and a wave forms off the coast of Japan, undersea sensors begin tracking its height, direction, and speed. This information pumps through computers and data banks that feed predictions of where, when, and how the wave will hit. Like the algorithms driving your social media feed, these predictions are only as good as the bank of experience they can draw on—the bigger the better. The warning system at the time relied on data banks that were devoid of anything like 2011. They struggled to simulate how the waves would interact with undersea formations, nuances of the rugged northern coastline, and any number of other imperceptible details. "We have to learn as we go and update our models as we can," Suppasri says. "It's difficult to define all the variables. For instance, some of the worst coastal damage might have been caused by a previously unknown undersea landslide." Such a thing hadn't factored into earlier models. It does now. Among the debris left behind in 2011 is a trove of new scientific data.

During the twenty to forty minutes between the shock and the first wave, sirens wailed and town criers sped through neighborhood streets blaring evacuation orders over loudspeakers. And yet many of those who died had been caught off guard. Thinking the worst had passed, some had returned to their homes after the earthquake. Others had never evacuated, thanks in part to early underestimates.

"I didn't even imagine that a tsunami was coming," said Toshiko Naganuma, whom I met on the Sendai plain. He had been a few miles inland when the ground started shaking but headed toward the coast to check on his house once it stopped. News alerts had first predicted a 9.8-foot wave. That was soon updated to 19.6 feet. Then, with the wave bearing down on the coast, the expected height was increased to 32.8 feet. Naganuma had just enough time to survey the damage to his house before it was torn in two and carried inland over a mile while he clung to the roof.

The 1970s-era warning system has since evolved. S-Net now issues its alert within three minutes of an earthquake and churns out increasingly accurate predictions as the underwater disturbance rolls through a more sensitive network of sensors. The models have also grown more complex and detailed, thanks, in large part, to all the additional reference points. It's still not perfect, Suppasri tells me, but it's as good as they get. However, updating the software was a breeze compared to the hardware.

The wave destroyed 118 miles of fortifications that had stretched for 186 miles along Tohoku's coast.[19] Afterward, the Ministry of Reconstruction had little confidence in the walls and breakers left standing and quickly set about designing new defenses. An expert at predicting tsunami behavior, Suppasri joined the effort. Out of the gate, there wasn't much of a discussion about *whether* to build new seawalls; it was taken as a given that people living in towns like Yuriage and Rikuzentakata, which were basically erased, would need better protection. Of all the ways to stop water, most look like walls.

Suppasri and the reconstruction team built silky smooth animated simulations that depicted how waves of various heights might surge through coastal towns. These advanced, three-dimensional models took into account things like distance from epicenter, beach slope, tidal stage, fluid dynamics, repeating waves, and seafloor terrain. Suppasri uses the word *bathymetry* like it's something people say all the time, because, of course, understanding seafloor topography is critical for predicting tsunamis. Waves lurch, dip, hole, swirl, run up, and climb depending on how water interacts with the ground beneath it. The wave that hit the wide and sloping Sendai plain peaked around 20 feet after advancing on oceanside farmland like a line of geese. Up north in cities like Ishinomaki and Kesennuma, it was funneled into narrow inlets with complex channels and steep beaches that pushed it up to 130 feet.[20]

One evening, while staying in a *ryokan* near Ishinomaki, I suffered a moment of clarity. The small hotel had an *onsen*—basically a communal hot tub where one bathes nude—that was set into the ground and encased in gray stone beside a patio garden with a bonsai tree and a small, trickling fountain. I sat with my back against the warm rock and felt my legs float up and my neck finally loosen in the hot mineral water. A small black beetle approached, and I watched it walk indifferently along the bath's edge just a few inches above the water line. As I tilted my head back, something tickled my nose. I lifted my arm to scratch it. When my hand plopped back into the water, it sent a ripple racing toward the bath's edge, where it smashed against the stone wall, splashed over the lip, and washed the beetle away in an instant. That's bathymetry. Immutable on any scale. I just sat and stared.

During the disaster, different seawalls in different towns had failed in different ways. Ishinomaki's flat-faced, 20-foot walls were reduced to pieces. Others crumpled in places or remained mostly upright and retained floods behind them. Concrete blocks 20 feet high and 16 feet long lifted up and slid.[21] Generally, researchers realized in hindsight that Japan's coastal fortifications had not been designed to be overtopped.[22] They were intended to hold their ground against seaward pressure, but as waves poured over them and scoured the land on their leeward side, their foundations weakened. The failure was compounded from decades of the walls doing exactly what they are supposed to do: blocking waves. Seawalls reflect wave energy rather than absorbing it and so erode at their base, eventually beginning to lean. The foundations were compromised, and it was the receding wave's backwash that often dealt the final blow by pulling the weakened blocks toward the ocean and unleashing the Pacific across Tohoku.

Suppasri's team understood that for the new walls to survive a larger wave, they would have to be more than just taller. Slanting eased wear on the seaward face and lessened the drag during backflow. Engineers settled on a slope ratio of two-to-one, meaning that a wall 30 feet high with a 10-foot-wide top would cover 130 feet at the base. Such a behemoth would no doubt change the face of any beach and promised to devastate fragile coastal ecosystems.[23] But how large would the new wall need to be? Should it protect from another 2011-sized wave? Something bigger? Or something more likely to happen soon?

When designing for the rebuild, the government relied on data that underestimated the Keicho Sanriku tsunami and categorized the 2011 wave as a once-in-a-millennium event. The prefecture divided its coastline into arbitrary zones and focused on protecting from respectable but reasonable waves that were likely to return within one hundred years. These would result from earthquakes of magnitude 7.4 to 8 and were classified as level 1 (L1). High-consequence but low-likelihood scenarios, such as 2011, were classified as L2 and were considered impossible to stop with seawalls, mostly because of the cost. Suppasri notes that L2 isn't even a true worst-case scenario. That would be a magnitude 9.5 earthquake ripping the ocean apart just south of Tokyo in the middle of a winter night at high tide. The government set the high bar at 2011, he says. "They had to move fast, but they have kept the worst of the worst in mind."

Other measures would be taken to insulate communities from future 2011s, but the immediate result of the simulated L1 waves was to dictate the heights of the new seawalls in each town. This became known as the L1/L2 scheme, and years later the phrase encapsulates a cultural universe in the same way as "her emails" or "the Immaculate Reception." Anyone paying attention knows exactly what it refers to.

~

In the summer after the disaster, prefecture officials presented their reconstruction plans to cities, towns, and hamlets up and down the coast.

"Reconstruction has been like a second tsunami," the survivor Naganuma told me. We stood in an expansive, grassy field that had once been his block. He held several laminated photos depicting his house before and after the wave. The entire neighborhood is gone now, relocated for those who chose to stay onto artificially raised ground 2.5 miles away. From this spot, "We can see the mountains now, but we didn't know they were there before, because there were houses all around," he said. We climbed a white concrete embankment that encased a small canal where he used to fish, and I saw that we were amid a table of farms and empty lots. Toward the ocean was a thick strip of short trees planted as a protective "green wall," and behind them was the wide gray line of the new seawall, which faded off into both horizons.

"We had so many houses, so many people living here, and it was all taken away," Naganuma said, holding the photos against a stiff breeze. "It's difficult to tell people how to protect yourself. The government didn't make enough effort to tell people about the history, and locals are hesitant to have this conversation."

There is one benefit to being a foreigner, I discover. Survivors are often willing to talk with me. I imagine that if I had lived through something like this, I would want to talk about it with others who had also, but 2011 survivors tend to avoid the topic with one another. I expect their reticence stems from a dutiful sense of politeness and deference, clichés of the Japanese character. Some say survivors fear upsetting others by raising difficult memories. "*Motto taihen na hito ga takusan iru*," they tell me. "Many others fared much worse."

In school, Japanese children are taught to keep their opinions to themselves, and this learned timidity played a crucial role in the reconstruction process. As adults, many watched quietly while someone else made decisions that would change their world forever.

Yorio Threatens to Leave

"I've been saying that if this seawall is built, then I will move out of this town. We don't need a seawall. I've been telling them that since the beginning," says Yorio Takahashi. He's one of the few who publicly challenged plans to build a 32-foot shoreline barrier in his home village, Ogatsu, and it ended up costing him everything.

Ogatsu is a hamlet in a narrow cove north of Ishinomaki. It's crammed on limited flat land between the sea and steeply rising mountains that hold uncommon veins of slate. Oral histories hold that locals have crafted this sedimentary rock into ink stones for 600 years. Takahashi, his father, and his grandfather had all carved and polished the slick black slate into rectangular wells used to hold traditional calligraphy ink, called *suzuri*. As one of the only places in Japan with deposits of the right sedimentary rock, Ogatsu is home to many of the country's best *suzuri* craftsmen, Takahashi among them.

Takahashi was at his workshop by the coast when the earthquake hit. He fled up a nearby hill and lingered before walking down to the beach, where he found three older men also milling around. Then, oddly, he noticed that the seabed was exposed. He could see the roots of trees twisting between rocks below the waterline and stranded fish flopping pathetically in the wet sand. White gulls swooped in to pick at them. When he looked up, he couldn't find a small island that had always been visible on the horizon. All he could see was the slate ocean.

Realizing what was happening, Takahashi turned to run. He yelled at the old men to do the same, but they called back that the water wouldn't go past their knees, so he grabbed the men by their sleeves and dragged them up the hill. They escaped just as the water came rushing in. Others had gathered in a wooded clearing atop the mound. Takahashi stood among them and watched his house snap into pieces. The house next door was also destroyed, with its inhabitants, a family of seven, inside. They had evacuated but then returned after the earthquake to retrieve *kaimyou*, mortuary tablets with the names of their dead ancestors. Atop the hill, Takahashi found a friend, and the two of them watched in horror as the black water spilled toward a hospital where the man's wife had gone to check on their two parents after the quake. Per official protocol, everyone in the hospital had fled to the roof. Takahashi, standing helplessly beside his friend, watched the wave overtake them all. They spent the night on the hill. At some point, the snow got heavy, and since they had burned through all the easily available deadfall, they added bamboo to a small pit fire. He remembers how it popped all night and kept them from sleeping.

He tells me this story while we sit in his tiny living room on coolers he uses for keeping caught fish. Its floors are tatami mats, and there's a small, low table between us, a small TV on the wall, and a couple calendars with fishing schedules. The adjoining room is empty except for some plastic boxes and a line of fishing rods. There's a slight smell of something rotten and a sense that he's just moved in, but he's been here for months. Not in Ogatsu. We're several hours' drive south in Fukushima Prefecture, because Takahashi kept his word: He left when the wall went up.

I've driven out of Sendai, down across abandoned coastal farmland and through the outer zone of fallout from the nuclear explosion that followed the wave on March 12. Driving through Fukushima, the organized chartreuse fields of Miyagi morphed into rough, weedy abandoned farmland. Gardens ran wild over gated back yards, which was something completely alien in manicured Japan. In Futaba, buildings along Main Street were crumpled and collapsed as if the quake had hit yesterday. Brick walls slumped off homes, exposing shoes, dinner sets, and sewing machines shrouded in a ghostly gray dust. Government officials had lifted the evacuation order after five years and were assuring the population that the area was safe from nuclear radiation, but people have been slow to return.

In his small apartment, Takahashi describes how officials came to Ogatsu that summer, waving designs for a 9.7-meter seawall. "We held meetings to discuss the plans, but the meetings didn't consider the peoples' ideas. They were just told what had already been decided," he says. This is something I hear in every town I visit. Many in Tohoku harbor a sense of quiet subjugation, a feeling of being left to wither while southern cities like Tokyo, Kyoto, and Osaka have bloomed. Japanese scholars I've talked with trace it back at least to the 1868 Meiji Restoration, when the Western-influenced, industrialized world finally dislodged the romantic image of samurai-led, feudal-age Japan.

It was a colonial-minded Meiji revolutionary named Kido Takayoshi who coined the term *Tohoku* for northeastern Honshu Island, as shorthand for *toi hokuteki*, meaning "land of eastern barbarians and northern savages." This prejudice dates back centuries. Power in Japan has always concentrated in the south, while the north has remained the "backward periphery ruled by the oldest form of feudalism," in the words of historian Ishimoda Shō.[24] After Japan reopened to outsiders and embarked on its own colonial exploits in China and Korea, Tohoku became an internal work colony, providing rice and labor but receiving little investment in return.[25]

That began to change, at least superficially, in the 1960s, after a devastating typhoon and a subsequent national investment in Tohoku on a previously unknown scale. This was during Japan's meteoric rise from defeated combatant to the world's foremost supplier of technical know-how and gadgetry, a period commonly referred to as the Bubble. Rallying behind fervent patriotism, the nation tapped into every source of economic potential on what little flat land it had, and it flourished, if only briefly. To support new industry, parts of Tohoku's coastal and riverside wetlands were drained and filled with earth. When that wasn't enough, entire neighborhoods were moved to make room for factories, processing plants, and oil refineries.

Building capital in vulnerable areas demanded a shift in how Japan regarded the natural hazard, from something to be evaded to something that could be prevented.[26] This concept, called *bosai*, underlies the construction of nearly 125 miles of coastal protective infrastructure: seawalls, breakers, and levees that were mostly intended to defend against king tides and storm surges.

They were "premised on the need to protect Japan's emerging capitalist infrastructure and cities from a seemingly capricious nature," writes the Chicago University anthropologist Michael Fisch. They also gave residents the confidence to construct communities right up against the water's edge and made nuclear power possible. "Locating such a machine on a tsunami-prone coast would be unthinkable if not for the belief that the tsunami as a natural event was entirely understood and predictable and that any serious damage could be absolutely prevented with the aid of a concrete seawall," Fisch explains.

In 1968, a tsunami rose to 20 feet as it hit Japan, and despite not being built for the purpose, the new walls turned out to be just the right height to stop it. To Japanese engineers, it signaled that they had overcome the problem. Just before 2011, an anthropologist named Ryan Sayre posed a question about tsunamis to one disaster expert who had written several books and had just given a lecture on the subject. She replied: "Tsunami preparedness is over."[27]

Afterward, when researchers revisited the relationship between existing seawalls and death rates in 2011, they found that the tallest walls were somewhat protective.[28] Smaller seawalls, however, those around 20 feet high, "showed no effectiveness in reducing impact." In fact, destruction and loss of life had peaked behind them.

~

The disaster is said to have shattered the public myth of safety on a scale not seen since World War II, but it also strengthened the bureaucratic framework underlying Japan's disaster prevention efforts. Miyagi's governor, Yoshihiro Murai, surveyed the damage that spring and declared, "There will be no more water coming into Miyagi!" After that, construction of a new round of walls was a done deal, Takahashi tells me.

One hundred fifty-eight people were killed in Ogatsu. Since then, the town's population of roughly 4,300 has plummeted to about 800 registered inhabitants.[29] Most of them remain only because they're too old to move. Young people and families have moved on. This hollowing out of communities is what had most hurt Naganuma on the Sendai plain. Without kids, he said, neighborhoods don't feel like home.

In the immediate aftermath, locals were generally supportive of new walls. It was only when the dust had settled that some began to quietly question aspects of the design or ask whether they should be built at all. That's what Takahashi did, and he feels as though he was giving a rare voice to a common idea. "I'm not against every seawall, but Ogastu didn't need the seawall," he tells me. The town sits in a narrow bay surrounded by green slopes. A paved road traces the waterline. "The people were already living up on the hill. Why not relocate homes farther uphill and elevate the road?" he suggested, but no one listened.

Japan's population began falling around 2010 and has been in a downward spiral ever since. The effects have been especially painful in Tohoku, where elders watch children leave for urban opportunities, taking any future promise with them. With fewer people to govern and a dwindling tax base, the region's cities have gobbled up administration of outlying smaller towns. As a result of this consolidation, quaint Ogatsu falls under the jurisdiction of Ishinomaki, a city in a different bay forty minutes away. Takahashi complains that Ogatsu's residents have little influence over decisions made there. It feels something like America's urban–rural divide.

The tsunami reconstruction plan came from the top and was to be implemented universally, said Hirotaka Abe, deputy secretary of Miyagi's Reconstruction Support and Memorial Division. We spoke in his tidy Sendai office. The national government dictated the terms and covered the costs, which were simply too great for the prefectures or cities to cover themselves. This put the national government firmly in the driver's seat, as the one picking up the bill often orders for the table. Prefectures submitted damage reports, and cities dealt with residents. Many communities had lost significant administrative staffing in the disaster, and so outsiders filled in to run local governments.

All along the coast, the plan was the same: Move neighborhoods uphill and build higher seawalls. "The government only needed some people in town to agree with plans for giant walls to move forward," Takahashi says. "So it sought out the opinions of those who had something to gain from it, and there were people who made huge profits from building giant seawalls." It's the old Japanese way, dating back to the early years of postwar growth: To save the economy, pour concrete.

Japan appears obsessed with concrete and cement. Long among the world's top producers, it generates an average of 6,104 million metric tons of the stuff each year.[30] Production peaked at just under 94 million tons in 1997, which is enough to make the Romans look like fiddling amateurs (although still a mere 3 percent of what China produces). Some of Japan's most powerful politicians, including a past prime minister, have stockpiled fortunes from the concrete industry. In the dim decades after its global defeat, bright, white concrete became symbolic of modernity.[31] Politicians who secured lucrative construction projects for their towns readily won reelection, and the government was often called *doken kokka*, a "construction state" that was—and still is—rife with corruption.[32]

In postwar Tohoku, the influx of concrete seemed proof that the national government was investing in their futures. Today, however, public sentiment has been tainted by the cement industry's role in driving the climate crisis: It accounts for 8 percent of the word's total CO_2 emissions, enough to make the cement industry, if it were a nation, the planet's third largest emitter.[33]

Concrete, the result of mixing cement with fine aggregates and water, is the second most consumed material in the world.[34] The only thing we use more of is water. How odd that concrete's role in muddling our relationship with water now has us applying so much of the former to contain, control, and convey the latter. Most of the product's greenhouse gas emissions derive from the creation of the cement that binds it. Limestone is baked at 2,600°F into what's called clinker, producing a carboniferous byproduct and burning through a hell of a lot of energy. Tack on the combustive outputs of material extraction, supply chain, use, and waste, this wonder material, so well suited to defense from man or nature, becomes its own problem. Of course, this has not slowed the global industry, which increased production by 275 percent from 2000 to 2021.[35]

In Japan, however, the years leading to 2011 saw the industry slip as population declined. Wealth drained from blue-collar Tohoku, and the region grew dependent on subsidies and massive public works projects, so that this reliance was well established by the time of the disaster. On one hand, plans to build new, gleaming white concrete walls in the aftermath looked like economic rebirth. But the companies profiting from their construction were often not based in Tohoku, and the new seawalls were no guarantee of safety. They are not tall enough to block another 2011-sized wave.

"The design of coastal embankments that are tall enough to prevent overtopping by the maximum feasible tsunami is financially impractical, and the effects of such tall structures, which would separate the fishing and tourism economies from the sea, are undesirable," Suppasri noted in a 2016 report. Under the L1/L2 scheme, seawalls stop lesser waves that are likely to return every century or so to protect investment in the low lands where farms and factories have risen in place of destroyed neighborhoods. When another big wave comes, they can only buy time for evacuation.

Still, despite the limited protection they offer, "The seawalls turned out to be higher than people imagined," Miyagi's Abe told me.

~

When Takahashi started asking questions at reconstruction meetings, officials invited university scientists who answered with jumbles of jargon. His ideas and suggestions were politely heard and then, as far as he could tell, forgotten. Prefecture officials had sent out carefully worded questionnaires immediately after the disaster and later used responses that supported beefed-up coastal protection to justify moving forward with seawall construction. But the months following March 2011 were a raw time when few were likely to argue with heightened protection.

Takahashi sat at the table during meetings in Ogatsu. He even brought alternatives to officials in Sendai. He put his opinion on the TV news for all to see. Eventually, a friend cautioned that if he kept asking questions, he would get himself killed. So he wrote out a will and then, in utter frustration, vowed to leave his hometown and abandon his inherited craft if the wall went forward as proposed. "There was no other way," he says.

Somehow, people in other towns had found other ways, though. Nearby Onagawa managed to cover their wall in grass to make it less of an eyesore on the landscape, and the fishing port Kesennuma had held out for less obtrusive designs. Two communities had even successfully fought walls altogether, but these small victories had been backed by wealthy and well-connected public figures. Little Ogatsu just didn't have the clout.

"So much of it came down to whose side their city administrators were on. In Ishinomaki, the city was with the prefecture. In Kesennuma, the city was with the people," Takahashi says, conjuring Kesennuma's name like a miracle. As a result, a white concrete wall now stretches more than 4 miles around Ogatsu Bay, encasing what's left of the sleepy hamlet. Most of the time it rises straight up, like a prison wall dwarfing a grown adult walking in its shadow. The thin river feeding the bay is also jacketed in sloping white concrete embankments, so that no part of the shoreline retains its natural state or a view of the ocean.

As in many other cities, Ogatsu's giant seawall doubles as a tourist center where visitors can learn about the disaster. On sunny summer afternoons, domestic tourists are content to shop for trinkets and children lick ice cream. Each of these centers brands itself around the local novelty, and Ogatsu's has an exhibit on the traditional ink stone making of Takahashi's family. It turns Takahashi's stomach. He comes off as subtly bitter but is adamant that he does not regret his decision. "There was no other way," he tells me again. Now, he works as a security guard and spends as much time as possible fishing from his small boat. He says he will be on the water tomorrow, and I wish him luck as I leave.

Exhausted and starving, I barely manage to find a rare *izakaya* open past eight o'clock. I order fried eggplant, seared beef, and a plate of buttery sashimi. I wash the hot and sticky day down with an icy glass of Asahi. Then I drown it all in sake, and I make a plan to visit Kesennuma.

A Fight Over Inches

The train from Sendai to Kesennuma takes a hair over two hours, but a speedy new coastal highway, the Sanriku Expressway, promises to shorten the distance to one hour and fifty-nine minutes, so I opt to drive. The road is good and I'm an American. As it turns out, the expressway is so new that stretches of it don't yet appear on the rental car's navigation system. Sometimes when I'm high on a brand new bridge, the navi urges me to turn, now, to plummet from a highway that it doesn't realize exists into an ocean I can see clearly many leagues down. For decades, this unbuilt Sanriku Expressway had been a symbol of broken promises to the people of Tohoku. Only when the government confronted the challenge of shuttling resources between coastal towns after the disaster did the road get built right quick.[36]

I speed north up the expansive Kitakami River valley and watch as bright white herons swoop over acres of chartreuse paddy. Rice is still raised throughout the country in these small parcels, mostly by old farmers who tend to them as secondary jobs. Japan is old—a quarter of the population is over sixty-five—and as in the United States, farming is a particularly aged profession.[37] Very few farmers have returned to fields in the inundation zones where homes, workshops, tools, and machinery were lost.

Climbing the mountains, I wash around inside white clouds that shroud the peaks in a thick mist, just like in those traditional silk paintings. The forest straddling the road is impenetrably thick with a kaleidoscope of green. It calls like Sirens for me to clamber out and explore, but when I come within a few yards the ominous trees send me retreating to the car. This is a far cry from the open, light-drenched ponderosa pinelands back home.

After World War II, Japan was caught in a nationalistic fervor of tree planting.[38] More than 40 percent of its native mixed conifer forests were replanted, primarily with Japanese cedar. Plantations are easy to spot today; they appear as huge swaths of conformity where, viewed from the right angle, trunks line up in neat rows like obedient soldiers. Cedar was chosen because it grows fast and is good for building. Unfortunately, the monoculture has since strangled biodiversity and increased the risk of landslides on neighborhoods below.

When the expressway veers north again, I catch glimpses of the coast through the V-shaped river valleys opening toward the sea. The seawalls are obvious. Rarely more than a mile away, they appear as stark gray strips between sky and earth in every valley. Those on the Tsuya River are particularly jaw dropping. Its mouth is fortified with a trapezoidal wall, and the jacketing continues about a mile and a half upstream under the highway to a concrete regulator. Lesser walls had protected the valley in 2011, but the wave traveled a half mile up the channel anyway. Now, the homes are gone and the Tsuya looks ready for an Allied invasion.

As a foreigner visiting Japan for the first time, I'm struck by how drastically this image differs from the fantasized depictions I've carried around for years. My notion of the country as a "forest civilization" full of robed monks bowing under twisted black pines is completely at odds with reality. To me, the Tsuya is proof of how impassively Japan has remade its landscape.

At last, Kesennuma comes into view. The city of about 60,000 people sits at the back of a narrow, northwest-trending bay. Commercial and industrial buildings cover a flat, low-lying peninsula at its center; neighborhoods sprawl up the western slope, and mountains rise steeply on the east. From a distance, Kesennuma gleams in the evening sun like Cloud City. A sparkling new suspension bridge with wishbone towers spans the bay. I wait for a hover car to zip across the sky, but this utopia is superficial only. When Takahashi said people here had been supported in their resistance to seawall plans, he meant only insomuch as they were able to win a few concessions. There are eighty-four seawalls in Kesennuma, more than in any other city in Tohoku. There are walls in the harbor and walls along the bay. There are straight walls, sloped walls, curved walls, and walls with windows. Concrete jackets 28 miles of the shore and riverbanks, and I watch as workers place more gray blocks.

As I descend into the center of town, my view in all directions becomes a monotony of slate. Concrete buildings, concrete foundations, concrete walls. The sky and mountains are overhead but not within reach, it seems, and I can't see the horizon. It occurs to me that I would have no idea if something were rushing up the bay right now. I feel like a peasant in some medieval castle under siege, anxious about what looms outside my confining fortifications. I find my hotel on the peninsula near the city's industrial core, drag my bag upstairs, buy an Asahi from a vending machine in the hallway, and call it a day.

It's morning and I'm on the hotel bed when it hits me. Eleven years, four months, and twenty-one days ago, this building had been alone amid a sea of debris. If you followed the events of 2011, you've seen Kesennuma, or what was left of it. A popular image circulating after the disaster showed a 330-ton blue and red shipping vessel perched like a discarded plastic toy atop piles of twisted metal just yards from this spot.[39] A sign bolted above the hotel's entrance shows the high-water mark. Thanks to its size, sturdy build, and no small amount of luck, this building survived, although it carries scars.

Last night, I was coming off a string of long days traveling in the oppressive summer heat. I had conducted hours-long interviews with people who had experienced more than I could hope to understand through a heroic interpreter who suffered from debilitating stomach pains. In my exhaustion, I had some trouble finding my hotel room. Misunderstanding the concierge, I took the elevator to the fifth floor and went searching for room 524. I walked up and down the halls, double checking every number until I found myself standing dumbfounded between rooms 523 and 525. I realized that there was no room 524. Rooms 514 and 504 were also missing, and there was nothing between the third and fifth floors. That's when I realized: There are no fours in this hotel. In Buddhism, four is the number of death.

With this in mind, I set out to visit Akihiko Sugawara, who runs a century-old sake brewery that narrowly survived the tsunami. He felt it was a miracle when he discovered the *moromi* still bubbling away in secure vats, and his team returned to brewing two days later. He also sits on the chairs of the city's slow food initiative and its reconstruction council. I'm sure he does more in a week than I do in a month, and that's with a personal brewery at hand. He tells me that if Kesennuma had any leverage during negotiations over its future, it was the port. The city is a home base for many global fishing businesses. It serves as a stopover for oceangoing vessels, providing mechanics and supplies, and maintains the facilities to process all kinds of seafood. Of Japan's 2,917 fishing ports, tiny Kesennuma ranks among the ten busiest.[40]

"It's a fishing town. Without the ocean, we cannot make a living," Sugawara explains from across a wide conference table. He points to a picture taken in the 1930s that shows wooden fishing boats docked in a row right up against markets on the downtown wharf. His building's narrow deco facade is there too; he salvaged the distinguished top half after 2011 and rebuilt the bottom to help preserve the waterfront's character. The disaster claimed almost everything around it, including 1,355 lives—among the most of any city—and about 40 percent of its homes, 80 percent of its businesses, and 95 percent of its fishery production capacity.[41]

Afterward, cities desperately needed money, and to get it they were required to submit a reconstruction proposal by that October, he says. So, "in the chaos of those early days, we produced a plan." It showed a strong preference among locals living close to the water for rebuilding without a seawall, and it even adopted a slogan: "Stay with the ocean."

"Locals are convinced that they can stay in close vicinity of the sea and yet defend themselves from natural disasters inasmuch as this relationship is maintained," he explains. "The larger issue is how to build better public safety infrastructure, infrastructuring the environment in a way that also allows for the preservation of an organic, everyday relationship with the landscape." Before 2011, "some bays already had walls to protect factories, but we didn't, and we didn't want one." They worried that a giant concrete wall would sever a crucial relationship between mountain, river and ocean, the people, and their livelihoods. "We had lived with the sea, and so we would keep living with the sea," Sugawara recounts.

The government's engineers brought their own plans, which, like those in Ogatsu, included a seawall as a precondition. They presented computer models and simulations and explained that, ultimately, the height of the wall would determine everything behind it, especially how close to shore people would be allowed to live and work.[42] In this way, the seawall became the gateway to the long process of rebuilding, and the initial offer was a wall 49 feet high. This single piece of infrastructure, which was just one part of an immense reconstruction effort, became the flashpoint for contention among neighbors who were still grieving an inconsolable loss. Caught over a barrel, they began considering designs. "If we had to move, this would no longer be our town," Sugawara says.

This idea of fishermen living in harmony with the ocean is not just some trite romanticization of coastal life. It's a relationship as deep and practical as the one between farmer and field or pilot and sky. Coastal communities have a multitude of tools for keeping the bond strong: folklore, ocean-facing festivals, shrines built to honor maritime gods. None are as important as sight. Without an open view, people can't see what's coming. This fact was apparent in Miyako, where video captured a woman riding a bicycle behind a seawall.[43] She did not know, because of the towering barrier that obstructed her view, that a roiling black wave was only seconds from spilling over the wall and washing her away. In those places, the walls were like blinders.

The view has less straightforward uses as well. "My maternal grandfather and my great-grandfather served as *daiboh*—heads of the local fishery co-op—for long periods of time," explained Hajime Chiba, an anthropologist in Kesennuma who was involved with the reconstruction discussions. "My grandfather would go out every morning to take a look at the sea, watching for the 'Kashima Current,' noting the color of the leaves of the big zelkova tree at Kumano Shrine at the edge of the sea in an attempt to predict the advent of a big catch of bluefin tuna. He had a lot of knowledge that I still don't understand."

And while his grandfather gambled out at sea, Chiba's grandmother worked the bounty of the beaches. Here, on this vital meeting of sea and land, she and the other women sold their catch, traded fish for whatever else the household needed, and collected clams, urchin, seaweed, and a number of other plants whose uses have largely been forgotten from the rocky shoreline's tidal pools. Chiba remembers tagging along with her and contributing what he collected to the family's dinner table.

"The issue of what to do about the beaches and rocky shoreline where seawalls are being built is about ecosystem services, which simultaneously involves the maternal function symbolized by these places' nurturing ability," he said. If we are really going to stay with the sea, "we must re-familiarize ourselves with the reasons for the traditional view of interaction and cooperation between the sea and the mountains."

Living in sync with the ocean also means noticing its subtleties, its changes, and its moods. Sanriku's tidal bays provide rich habitat where fish and shellfish breed, lay eggs, and mature safely outside the harsh open sea. Young marble-sized flatfish hide in kelp forests, where they grow large enough to venture from shore and eventually lure bigger migratory fish like salmon and tuna.[44] Over the past fifteen years, Japan's salmon catch has taken a 70 percent dive, largely because of changes in the Pacific's currents spurred by melting Arctic sea ice. Fishermen now hunt for fish farther from shore than ever before, at an unsustainable cost, and the additional impacts of new seawalls might not become clear for decades.

Before driving to Kesennuma, I went to the beach with Noriko Uchida, an ecologist at Tohoku University who worked previously at the Ministry of Environment. We walked along a narrow strip of sand that was left between a sloping wall and the Pacific's lapping waves. In the sand all around us were hundreds, thousands of small red stones, all flat with rounded edges, that looked like sedimentary rock. Picking one up, I saw that it was imprinted with a number and realized that these were not stones but pieces of roofing tiles from homes washed out to sea.

Uchida specializes in using DNA analysis of water samples to track plant and animal life. She says the method, called environmental DNA, is improving and is becoming popular at cash-strapped government agencies that see it as an alternative to field observation, but environmental DNA still provides only part of the picture. It can tell her that a fish is in the water, for instance, but not whether the fish is healthy or infected or old or an odd color. That level of detail requires spending time at the water.

On the surface, the wave devastated coastal ecosystems, but scratch a little deeper and the story changes. Uchida was amazed at how fast species rebounded. Some that spring up just once every ten years took advantage of the disturbance and sprouted early. Others reclaimed their places quickly. In tidal flats across Sendai Bay, research conducted with the help of citizen volunteers revealed that ecosystem changes decreased from year to year after the tsunami and largely disappeared by 2018.[45] The natural world is more resilient than we give it credit for. But then came the levees, seawalls, and engineered coastlines.

"For nature, reconstruction has been like a second tsunami," Uchida explained. Land was elevated, burying everything that lay beneath it, and productive wetland was filled with soil from nearby mountains that carried alien species. Concrete was poured, barriers were raised, and the coast was altered forever. Ecological surveys show that these efforts have devastated some sea life, particularly species of mollusks, some of which have been wiped out.[46] The risk is not unique to Japan. Armoring coastlines encroaches on natural habitat the world over. Seawalls and levees on rivers emptying into the ocean simplify complex waterscapes. They increase turbidity and spoil intertidal zones with a deluge of nutrient- or chemical-rich runoff.[47]

In studies of coastal armoring in California and Chile, placement mattered.[48] Wildlife that made its living close to the barriers fared the worst. Sand hoppers—the Talitridae that spring from your approaching feet on beach walks—were harmed significantly. They may not seem like much, but they indicate shoreline changes above their weight class. Coastal biodiversity takes an even greater hit when seawalls come in regular contact with waves and tides and reflect their energy. This drives unequal erosion, which messes with natural nutrient mixing, and shifts sediment loads. Sea-level rise will only worsen this problem: Seawalls don't leave any room for beaches to go as the water rises. As a result, they'll disappear rather than retreating, and salty waves will crash against concrete buttresses instead of sandy slopes.

In Tohoku, fishermen described seaweed forests morphing and moving and lichen changing color. Chiba said the abalone catch in his local bay has declined by a third. Uchida wonders whether bees and butterflies will be willing to cross a wall's expanse to reach the nectar of lilies and marigolds outside the fortifications.

"Biodiversity monitoring and surveys take time and money, so sometimes it isn't done," Uchida told me. And sometimes companies conceal the discovery of a rare or threatened species in their work zones "because that will stop construction."

Once the wall is in place, the local environment is fixed for the life of the infrastructure. Exposed to the ocean's constant battery, concrete seawalls may begin to break down within fifty years, but they are intended to be maintained for much longer. However, much of Tohoku's existing infrastructure built during the postwar Bubble is already aging out with its people. This raises a pressing question: Who will pay to maintain these structures once the population and tax base evaporate? The country depends heavily on tourism, but Tohoku's business owners also wonder who will drive all the way just to stare at concrete. Desperate to restock rural towns, Japan is paying Tokyo residents to move to the country and encouraging young people to drink more, hoping that booze will grease the wheels of procreation.

It's encouraging, at least, that beaches have been shown to rebound when allowed. In Washington State's Puget Sound, for instance, the insects, crustaceans, worms, snails, and clams that feed Chinook salmon were back on the menu less than a year after concrete armoring was removed and habitat restored.[49] From her time at the Ministry of Environment, Uchida knows "green" or natural solutions, such as giving rivers room to flood or substituting sand dunes for concrete barriers, are also earning respect. "Local governments want to save their money and have to consider where to cut costs," she said. Once established, natural green infrastructure is nearly maintenance-free, and the ministry is paying attention. It even has a committee to study the potential, but these methods are not common. "I'm Japanese, but sometimes I wonder why it's so difficult for Japanese people to change. The culture is very resistant to change," she told me as we left the beach.

~

During talks about Kesennuma's future, old ways emerged in classic forms: an unshakable obsession with technology and what Chiba called "disaster paternalism." Prefecture officials came to meetings with mathematical models and diagrams. "They told us we are amateurs," he said. "We understand that the government should try to save our lives, but they cannot understand the life and work of a fisherman. Japanese people accept their position. They know it's dangerous, but they also know it's a blessing. They stand up again in the same place and start over."

The negotiations dragged on. The prefecture had its marching orders from the national government, but Kesennuma had a prominent fishing port and the leadership of people like Sugawara. In the end, all they wanted was to be able to see over the wall, Sugawara tells me at his harbor brewery. So the government tweaked its simulations by making small variations, trying to trim just 5 or 10 centimeters off the planned wall so a person walking past could glimpse the sea. These tweaks had tradeoffs. If they built a breaker out in the bay, it might slow a wave enough to shave off a meter or so, but that would also disturb natural circulation and affect oyster fishermen. If they moved homes farther back, they could shave another meter. If they raised the land, the wall wouldn't seem so high from the landward side, but the city would have to pay for that itself.[50]

After five years and at least one hundred meetings, Kesennuma finally struck a deal: It would elevate the land by 11 feet and build a 13.5-foot seawall. The top of the wall would have a 3.2-foot metal flap gate that would lie flat until a wave automatically triggered it to swing into place, making the barrier 1 meter higher. Through this mathematical ninjutsu, the wall would reach a total height of 16.7 feet above sea level, and when the flap was down there would be just enough space for a person walking along the concrete boardwalk behind it to see over. Everyone was relatively happy—that is, until the isostatic rebound.

When the Pacific plate slipped, causing the earthquake, the ground under Kesennuma dropped 70 centimeters, and for some reason it rebounded faster than anyone expected. Four years into constructing the wall, the land had risen by 22 centimeters, which rendered all the fretting and arguing over 5 or 10 centimeters pointless.

Memory in Bloom

I can tell that some of those I speak with wish the topic of seawalls would simply go away. I travel around Tohoku talking with people who lost everything, asking that they bring up the memory for my benefit. I tell myself this is valuable work, that there's something to be learned here and that those of us who can must spread the word. But deep down I feel as if some uncaring force is having its way with us. I don't mean god or even politics. I mean a system so ruthlessly efficient in its controlling of our lives that it could only have evolved along with us, as part of us, to become as resistant to change as we are.

Chiba and Takahashi aren't just asking that people be allowed to decide how their neighborhood should look or what risks they're willing to endure; they're challenging a system of apparently boundless capital and technical expertise, which has taken us so far so quickly on the backs of so many miracles, and asking whether that system might be maladapted to its primary objective of protecting our wellbeing. When the Japanese government made seawalls the gateway to reconstruction, it made seawalls the manifestation of this discord and monuments to a deep rift that goes far beyond the shores of Tohoku.

"Miyagi Prefecture promoted the seawalls because, even if people reject them now, they might be valued in a hundred years when another big waves comes," said Satoru Imakawa, a journalist and Kesennuma city council member. "But the local people reject that logic, saying they want to pass down an intact natural world and its beautiful views to future generations. Both groups are looking one hundred years down the road and arriving at different places."

And neither wants to do something they end up regretting.

So where are the kids in this story? They'll be the first to live behind these structures, untethered, some say, from the ocean that supported their ancestors. Like the first generation to live in a world where cell phones are ubiquitous, they'll know nothing but the effects of our decisions. "Young people don't want to live here," said Imakawa, a father of three. About 80 percent of them leave Kesennuma after high school and never return. Even before 2011, they were leaving for better jobs down in Tokyo, Kyoto, and Osaka, and the disaster has sped the drain along. While the country's birthrate has fallen by 30 percent as a whole, it has plummeted by 70 percent in Kesennuma. All of Tohoku is desperate for young people.

The walls were intended to lure the kids back. Concrete conveyed a sense of progress and safety to a generation. It was hoped that it could help restore Tohoku, but officials in cities hours away by car or bullet train might have miscalculated. "If the disaster had happened thirty years from now, there probably wouldn't be any seawalls," Imakawa said. "The kids who are growing up with the decisions made today will be in charge later. The people moving in now want nature."

I leave Kesennuma and drive to Rikuzentakata to meet with one of those young people. Like Ishinomaki and Kesennuma, Rikuzentakata is also vulnerably situated in a southeast-facing bay and was hit hard in 2011. Unlike in Kesennuma, where limited flat land forced people to build their homes uphill, Rikuzentakata's buildings were spread across a low plain, and therefore the city wasn't so much devastated as erased. I approach from the south on the Sanriku Expressway, streaming past one of the most affecting images of the disaster.

The site of a historic tree-planting effort, Rikuzentakata's beaches were lined with at least 60,000 black pines on March 10, 2011. When the sun rose on March 12, only one remained. This "Miracle Pine" later died because of the residual salinity, but the prefecture immortalized it, artificially, as an 89-foot-tall reminder. It casts a spindly shadow over a new memorial center filled with disturbing images of the event. Behind the fossilized tree stands a crumpled two-story building that used to be a youth hostel. It has also been left in place, with its north end slumped over and its sloughed walls exposing its abandoned guts. It's a ruin, left as it appeared that day so that all who drive by are reminded. Ruins dot Tohoku: hospitals, high schools, apartment buildings, all uninhabitable monuments to death. The discolored waterline is clear on the walls of their upper floors, their ceilings hang from their rafters, and twisted rebar pokes from their exposed frames. Behind the preserved building is a new 41-foot-high concrete seawall that connects with a concrete floodgate spanning the newly fortified Kesen River.[51]

There had always been a seawall in Rikuzentaka, says Shoma Okamoto, so the new one wasn't a surprise. What's surprising is how few people are willing to live behind it. Okamoto is one of those badly needed young people who has returned home to Tohoku with renewed purpose and fresh ideas. While the government has elevated the land, raised the seawall, and built a massive lift gate across the river, he's been planting thousands of cherry trees, *sakura*, in a line that traces the wave's high-water mark.[52] When they bloom in spring, he hopes the explosion of pink and white blossoms will provide a warning as impactful but more beautiful than the decrepit ruins.

Okamoto was a young man beginning a promising career in architecture before 2011, and he had flown from this blue-collar port town for what seemed like greener pastures in Tokyo. He was away when his family home was obliterated, and he returned three days later. "I had a feeling of guilt as a survivor," he tells me. "I hadn't been here to help when it happened. I wasn't doing anything for the community."

We walk near the base of a hill near a large Shinto temple. Behind us, the land spreads out flatly toward the ocean, which we cannot see because of the wall. There is a smattering of newly built, prefabricated apartments nearby. Blue ridges hem us in to north and east, forest grows thick up this hill, and unkempt grass fills open lots all around. It's another hot afternoon. I can feel the sweat soaking through the back of my shirt. Somehow, Okamoto is dressed in black slacks, a navy blue shirt, and black leather shoes. He wears round black glasses and black stone earrings. His black hair is long on one side and cut at an angle, a style popular with Japanese boy bands. I can tell he's spent time in Tokyo. He speaks quickly and punctuates his comments with a hearty "Hai!"

He says being wiped off the map might have provided an opportunity for Rikuzentakata to build back better, but instead the city is being rebuilt to its pre-2011 capacity. The civic center will hold the same number of spectators as the old one. The power lines were restrung above ground rather than being buried, as so many residents requested. He sees this as a missed opportunity to improve, and it's taken a visible toll.

"The government's reconstruction plan had a deep structural problem," he says. The existing seawall was more than doubled in height, and the land behind it was elevated by 34 feet, paid for by a special reconstruction tax. At the start, the prefecture asked how many people would choose to return to the elevated land if they could and then raised enough lots to accommodate them. However, the first homes did not become available until 2017, and in those excruciating six years, most people left. Because tax money was used to elevate the land, residents are not allowed to sell their unused parcels, and so half of it sits idly in weed-covered lots.

We climb the slope behind the temple to where his group, Sakura Line 311, planted its first trees on November 6, 2011. Below us is a grassy field where there was once a kindergarten with sixty kids. The earthquake struck just as parents began arriving to pick up their children, and government rules stipulated that the school must release the kids, even if the teachers knew doing so might doom them. The children whose parents hadn't yet arrived survived by fleeing up this hill with their teachers.

"My motivation is to not let something like this happen again," Okamoto says, and to that end, his team has so far planted 1,918 trees. They use 600 varieties of cherry, and although some have been bred for their height or exceptional coloring, he prefers hearty native strains that need less upkeep. Sakura are a novel take on an age-old custom. After past disasters, generations of Japanese people built shrines and erected stone obelisks to pass along their story and caution against rebuilding on inundated lands. These *fukkouchis* exist all along the coast today, but Okamoto says their meaning has been lost. "They are cold, like tombstones. Not a positive feeling," he says. Also, the stones don't need to be maintained, and so once carved and placed, they're easily forgotten where they stand, covered in leaves and moss on forested slopes. Growing up in Rikuzentakata, "I didn't know about these warnings," he says. This is something he regrets.

The Sakura Line's motto is "We regret." By that, Okamoto means they regret what happened, who died, and how. But they also regret having lost the memory of previous disasters that might have saved them, and they regret having put so much of their faith instead in technology that didn't. "We lived that moment; it was a significant time. It is important that we find something to learn from it," he says now. "Disaster prevention with physical objects was not enough. There needs to be a psychological option."

I heard something similar from Ichiyo Kanno, who runs a *ryokan* in a small bayside town farther north. She had lost her husband, daughter, and son-in-law all at once—not in the tsunami but years later in a fishing accident. I asked why she remains in the disaster's wake, in the same town, the same house, with all those painful memories. She replied that she and the house had a special bond, which she called "*En*," something akin to karma or fate.

"It's here that my husband and I got married," she said, sitting on tatami mats in the large wooden house that she had converted into a bed and breakfast after 2011. “It’s here that I’m attached to. It has our souls in it. It has to be here.” *En* was finding meaning in random moments or unintentional happenings, even bad ones. After the disaster, volunteers came from all over the world to help Tohoku rebuild. She opened her once-empty home to them, and they gave her renewed purpose. We spoke over the clatter of a construction crew banging together a new seawall a few yards away through the open window. Ishiyo said the wall will not be tall enough to have protected her before and isn’t going to make her leave now.

In Rikuzentakata, Okamoto knows a thin line of cherry trees won’t stop a tsunami, and he admits that aside from the few weeks in spring when they bloom, sakura look much like any other tree on the hill. But so far, hundreds of school children who have no memory of 2011 have knelt down in the soil beside people who survived the day and the days after, and on those afternoons stories were shared. Over twenty years, he intends to plant 17,000 cherry trees, which he estimates will oblige the help of 50,000 people. Afterward, the trees will need to be trimmed. Someone will have to prune their branches and protect them from hungry insects and parasitic fungi. A generation from now, when someone with hillside acres through which the line of planted trees passes decides to sell their home, the realtor will point out the sakura and explain where they came from. And for a short time every April, a line of pearl, crimson, and fuchsia will spool out, encircling the city in a striking thread of remembrance.

On those days, Okamoto hopes his answer to the government’s concrete will change how people view Rikuzentakata, from a place of unmeasurable sadness and destruction to a vibrant town in full bloom. “Fifty or one hundred years from now, people who didn’t experience 2011 will come and wonder why there are so many sakura trees in Rikuzentakata. It will start the conversation,” he says. But first, “We have to admit it happened, get over our regret and move forward.”

~

Around here, the only thing worse than remembering is forgetting. Tohoku is littered with memorials, museums, and ruins. They are their own form of adaptation that telegraphs a key lesson from 2011: Some threats can't be stopped; they can be either avoided or endured. The simplest reminder of this fact is often the most effective.

Driving south through Sanriku's rugged terrain on my way back to Sendai, I am constantly passing beneath blue signs bolted to traffic signals that indicate when I'm dipping into the inundation zone. This is the area of sloping coastline that flooded in the past and is likely to again. Some signs list the elevation—8.9 meters, 3.0 meters, 7.4 meters above sea level—which, when compared to the 40-meter wave, is a tad disconcerting. At each sign, I gaze outside my window and imagine it's that day and I'm speeding along toward higher ground, with a roiling black sea engulfing everything behind me.

The coastal highway leads me past emerald forests thick with pine and cedar blanketing the way to land's end. In the largest example of postdisaster green infrastructure, more than 110 square miles of this ragged coastline was collected under the umbrella of the Sanriku Fukkō National Park.[53] *Fukkō* means "reconstruction," and in this case it opened the door to large-scale environmental conservation. The park's dense forests, pebble beaches, quiet coves, wetlands, and grasslands shelter endemic species of flowers, enormous sea eagles, buzzing salmon runs, and the lightning-fast deerlike Japanese serow. In a country replete with historic monuments and cultural attractions, the park represents something unique to Tohoku.

The park also capitalizes on the rising tide of outdoor enthusiasts, often young people who have grown tired of crowded city life. They wear fleece, carry backpacks, and drive Subarus, and like their American counterparts, they're an economic opportunity—in this case, one that could bolster a tax base saddled with the cost of maintaining enormous adaptive infrastructure. The park intends to draw them in with its fresh air, gorgeous views, and swimming holes. There's even a 750-mile trail—the first long-distance path in Japan—that offers Japanese hikers something like and yet entirely unlike the Appalachian Trail experience. As with Okamoto's line of sakura, the park will not stop an incoming wave, but it does show how the country's relationship with nature may be shifting as a result of disaster, and that's inspiring nonetheless.

Farther south, green views give way to developed suburbs and Sendai's cityscape. I continue past the city and across the lower plains until I reach Iwanuma, where Naganuma rode out the wave from his floating roof. Here, I pull over at one of the most controversial examples of Japan's capitulation to changing tastes. The Millennium Hope Hills are a unique adaptive project that encompasses fourteen square hills like ancient Mayan pyramids, made from collected piles of disaster rubble 36 feet high and covered in grass.[54] They are evacuation sites. Around them, where there had been vibrant neighborhoods and busy markets, are now solar farms, industrial warehouses, and recycling centers. These are facilities Japan is willing to risk losing in another large wave. That's the kind of decision that gets made when the worst-case scenario becomes reality.

There is also a new seawall, of course, and so I climb one of the hills for a view of the ocean behind it. Cresting the top, I look west past a north–south highway that has been elevated as a secondary inland barricade to dark mountains; then north to the Sendai airport, which was engulfed in 2011; and east toward home, a long way out. Between my hill and the seawall is a strip of densely clustered young trees. This is the Great Forest Wall, a defensive green line modeled somewhat after the pines Date Masamune planted to shield his seventeenth-century fiefdom. Forests, which can generally withstand a 16-foot tsunami, played a role in protecting cities like Ishinomaki in 2011.[55] Phalanxes of trees reduced the wave's destructive power and delayed its arrival time by six invaluable minutes. The forest's greatest benefit is unsettling, however: keeping things like houses, cars, and people from washing *out* to sea.

I stand under a ramada like you'd see in any American park, except that its pillars are hollow and filled with dried food, blankets, and first-aid supplies. Its benches convert into stoves, and it has shades that pull down to keep out the weather. Taken together with a moving memorial center, the evacuation hills, forest, elevated roads, and seawall combine to form a multilayered defensive fortress that is both gray and green. It's a welcome reprieve from the brutalist, wall-dominated scenes I've been touring, but I still wonder: Who is it protecting? There's no one around.

The Hills project drew controversy for its cost, because its forest wall relies on nonnative trees, and because it's not clear who would need to evacuate, because most of the people have been forced out. Still, it found a champion in one of Japan's leading tsunami scientists, Fumihiko Imamura, director of IRIDeS. On a Friday during the summer break, when students are away and Sendai is gearing up to a weekend of festivals, I walk across campus to the IRIDeS, pass a TV in the lobby that plays a continuously looping simulation of the wave hitting Kesennuma, climb the stairs, and find this jovial sage in his quiet office.

In the 1980s, Imamura had worked on some of the earliest mathematical schemes for predicting inundation from tsunamis. These equations factored variables like earthquake severity, crustal slippage, undersea topography, beach slope, elevation, the list goes on, into eight-bit simulations that almost brought the math to life. Although they looked like a game played on a Sega Genesis, these early models put Japan ahead of the curve and made Imamura a world-renowned expert. They almost thought they had tsunami preparedness figured out, he tells me, and then came 2011.

"We were shocked. We didn't have a system to consider such a large and complex event," he says. Others would describe what happened on that snowy March day as something too large and too complex to be understood, but IRIDeS was born the following year with the determination to overcome existing ignorance. It has since grown into one of the world's foremost laboratories for the study of tsunamis, where scientists like Anawat Suppasri, Yuichi Ebina, and Noriko Uchida tackle the problem from myriad angles. Even humble and soft-spoken Imamura has become something of a national celebrity.

After the disaster, Japan turned its gaze to the horrors unfolding in the north, where Imamura's visage was a common sight. He was captured on television, profiled in magazines, and quoted in newspapers. When it came time for the government to present its plan for building seawalls up and down the coast of Tohoku, Imamura was their expert tsunami engineer. He was one of the guys with the charts and the jargon. He talked with survivors in devastated coastal towns and sat through an onslaught of heated arguments over ideas. It was his role to bring the sterile scientific perspective to the table, and in the early days, he says, they appreciated him for that. After all, he was among the scientists who developed the infamous L1/L2 scheme.

Despite how it seems now, that framework initially represented a significant shift in Japanese thinking about disaster. By creating the level 2 distinction, which encompassed waves no seawall could contain, the country's engineers allowed for the possibility that they could not control everything nature might throw at them. They relented.

Saving lives was the priority, Imamura tells me. "In order to get agreement from the people and the government, we needed a consensus over the heights of tsunami countermeasures. So, we divided into two levels. In level 1, we would utilize structural measures such as seawalls to protect lives as well as the community itself, factories, industries, ports—everything we needed to protect. But sometimes the tsunami will be higher than this level, and we cannot stop them all." So, his team plugged its numbers and ran its models, and the plan it came up with was the plan the government adopted to tackle an impossible task, the plan that became standard issue throughout eastern Tohoku, the plan that set neighbors at each other's throats over a measly 5 centimeters—the plan that came to change the face of Tohoku forever.

"In the first year after the disaster, people wanted the safety of walls," Imamura tells me. They were in favor of construction. But then a year passed, and then another, and their minds began to change. The incapacitating waves of terror spurred by recollections of the day had faded some, and although they would never fully slip away, they became folded into an entirely new context. Nothing would change what had happened, and so people focused on what would happen next. As each meeting adjourned with more jargon than discourse, simple people in small towns felt their grip loosen on that one small slice of control.

The national government had set out to rebuild Tohoku within ten years, and this was a mark it would not miss, but the task swelled to an enormity rivaling the disaster itself. Every little hamlet and bay village had a special request, an exceptional circumstance, an idea for redesigning the wheel, and so officials ushered the proceedings along, hoping to still make their target. "Perhaps the time set aside for community planning had been too short," Imamura admits now.

After watching communities unravel around the seawalls, Imamura says his team came up with up a radical concession, an adaptive plan that would give locals time to consider what was being built. In this new scheme, crews would build walls about half as high as planned and leave them in place so people could experience living behind them. After a few years, if they still believed giant seawalls were their best bet for protection, the walls would be augmented. Imamura's IRIDeS team had even engineered a way to ensure the walls would be structurally sound at both half or full heights. However, if those who had experienced the absolute worst nature could throw at them were still unconvinced, the walls would be left a mere 16.4 feet tall. The plan never got any traction, though. Imamura says the Ministry of Finance quashed it for budgetary concerns. Japan had imposed a special recovery tax with a ten-year duration, and there was no political certainty that funds would be available once the decade was out. So they used the best science available at the time and built the walls as quickly as possible. No one thought it would be perfect.

As always, it came down to time: a decade to meet the official reconstruction goal. Just a few months to decide what kind of walls each town would choose to live with forever. Too many years spent arguing before new homes were built. More than a decade clinging to the memory. Four centuries since Date Masamune's samurai reported the great wave they had weathered off Sendai. Of all the lessons Japan's experience has to offer a world confronting climate change, the dubious role of time might be the most important.

Without a plan in place for reacting to an event that seemed too distant, the country was left scrambling. Initial intentions to build back better fell prey to the ticking clock and devolved into build *something*. Opportunities were lost. People left. In the end, those who remain are saddled with a middle ground of fortifications that would not have protected them from 2011 and might not protect them from lesser waves as the seas rise.

Despite being built to fend off threats on the scale of centuries, reconstruction designs and risk assessments do not account for the impacts of climate change–driven sea-level rise. In a research paper from 2019, Reconstruction Agency officials explained this omission to a team of researchers from the University of Sheffield, saying, "Sea level is only rising very slowly."[56] The paper's lead author, Peter Matanle, compared the omission to the French Maginot Line of the 1930s, which was built to defend against a repeat of Germany's assault during World War I but ended up allowing the occupation of France. Estimates of sea-level rise are constantly being updated, but the change appears to be accelerating. Some experts predict as much as a 5-foot increase by 2100. This could be enough to bump a manageable L1-sized wave into L2 disaster territory, for which the coast is unprepared.

In April 2011, then–prime minister Naoto Kan proposed a reconstruction of Tohoku that, rather than being "stuck within a traditional framework," would prioritize local ownership, creative outcomes, and "harmony among nature, human beings and technology."[57] He was forced to resign five months later. Instead of true transformational adaptation of its coast, Matanle writes, "the Construction State and its logic of modern developmentalism as a defensive bulwark against the assumed destructive encroachments of nature remains in place."

Cities such as New York and Miami, which are spending hundreds of millions of dollars to fortify and armor their shores with protections that will be no match for the rising seas in fifty years, should take note.[58]

While Imamura explains what happened in the aftermath of the tsunami, Sendai prepares to kick off its annual Tanabata Festival with an elaborate fireworks show. Imamura says we should remember what happened to the old samurai, Date Masamune, and I'm a bit surprised when yet another conversation comes back to this old *daimyo*. But Imamura describes the seventeenth-century samurai as a civil engineer with remarkable foresight and a keen sense of adaptation. He planted his own green wall, dug a canal that eventually slowed the 2011 wave, and erected kilns to make use of the salt that the 1611 tsunami left behind. After what historian Ebina would call "Keicho Oshu," Masamune also appears to have pulled his cities back from the inundated areas. Perhaps most importantly, the venerable warlord tried to gather and pass along stories of the last great wave. Tragically, it was forgotten.

"We need to transfer the information through future generations," Imamura says. "People forget the meaning of stones and pine trees." The story must be threaded through everyday life.

This is a great challenge for Japan and, I believe, for all of us: How do we ensure that our adaptations travel through time with their meanings intact? How do we avoid the sense of urgency we build them with from fading into complacency over centuries to come? Because we're talking about centuries. The effects of climate change might feel immediate today, but they will last and escalate for generations. In one hundred years, how will people living along the Sendai plain view those fourteen strange hills and the dense strip of forest by the shore? What about in two hundred or four hundred years, when, as Ebina believes, they are likely to experience another great tsunami? Will they have returned to the sea, using the hills for some unimaginable new purpose and living in the shadow of huge walls? Will they remember that the walls are not tall enough to protect them?

It's Friday evening and Imamura is eager to get home, so I thank him and we bow for what feels like an eternity. Then I depart, gifting him a decent bottle of sake and walk east toward Masamune's castle. Its ruined black walls and uncovered layout are now a popular historic site. There's a monumental bronze statue of the lord on his horse, the remains of a stone tower, and, of course, a gift shop. The castle is perched high, with a panoramic view of Sendai spreading out before the Pacific Ocean, and I decide to view the festival's fireworks from its darkly mysterious grounds. Plenty of other people have the same idea. Couples intertwine on blankets rolled across the grass, kids run back and forth with mouths full of candy, and amateur photographers set tripods in front of everyone else.

Today, the Tanabata festival is caught up in the story of two heavenly lovers separated by the Milky Way. People dress in summer kimono and write wishes on strips of paper that they hang from elaborate decorations. Long ago, it was a time to welcome the spirits of the dead and remember how they came to pass. This communal habit is crucial to keeping the living in communication with the dead, their stories and lessons, but like summer festivals everywhere, that meaning has evolved. Eventually, it came to mark a time to cut loose, drink cold beer, and eat fried meat on a stick beneath fireworks. Someone told me that after 2011 people remembered Tanabata's older purpose; it took a year like 2011 to remind them.

We talk about past disasters with an eerie reverence. The Great Chicago Fire. The *Titanic*. The Johnstown Flood. Hurricane Katrina. March 11, 2011, became one of those, tragically enshrined in our collective memory, and these people lived through it. I try to keep that fact in mind as I ask them to relive it. In a way, though, I'm living through one, too. We all are. A great and undiscerning wave is overtaking us all, like a tsunami in slow motion. Those who survive will pass on stories of its advance to generations who will decide how to mark the remembrance and what lessons to glean.

As the sky darkens at Masamune's castle, Sendai's modern cityscape twinkles behind the ominous shapes of its ancient battlements. Anticipation grips the crowd. Parents point out across the river, directing their children's eyes to where the fireworks will appear. Then a sudden cackle of staccato bursts announces the beginning of the show. The sky cracks wide open, and all around me faces are bathed in the glow of an ageless light.

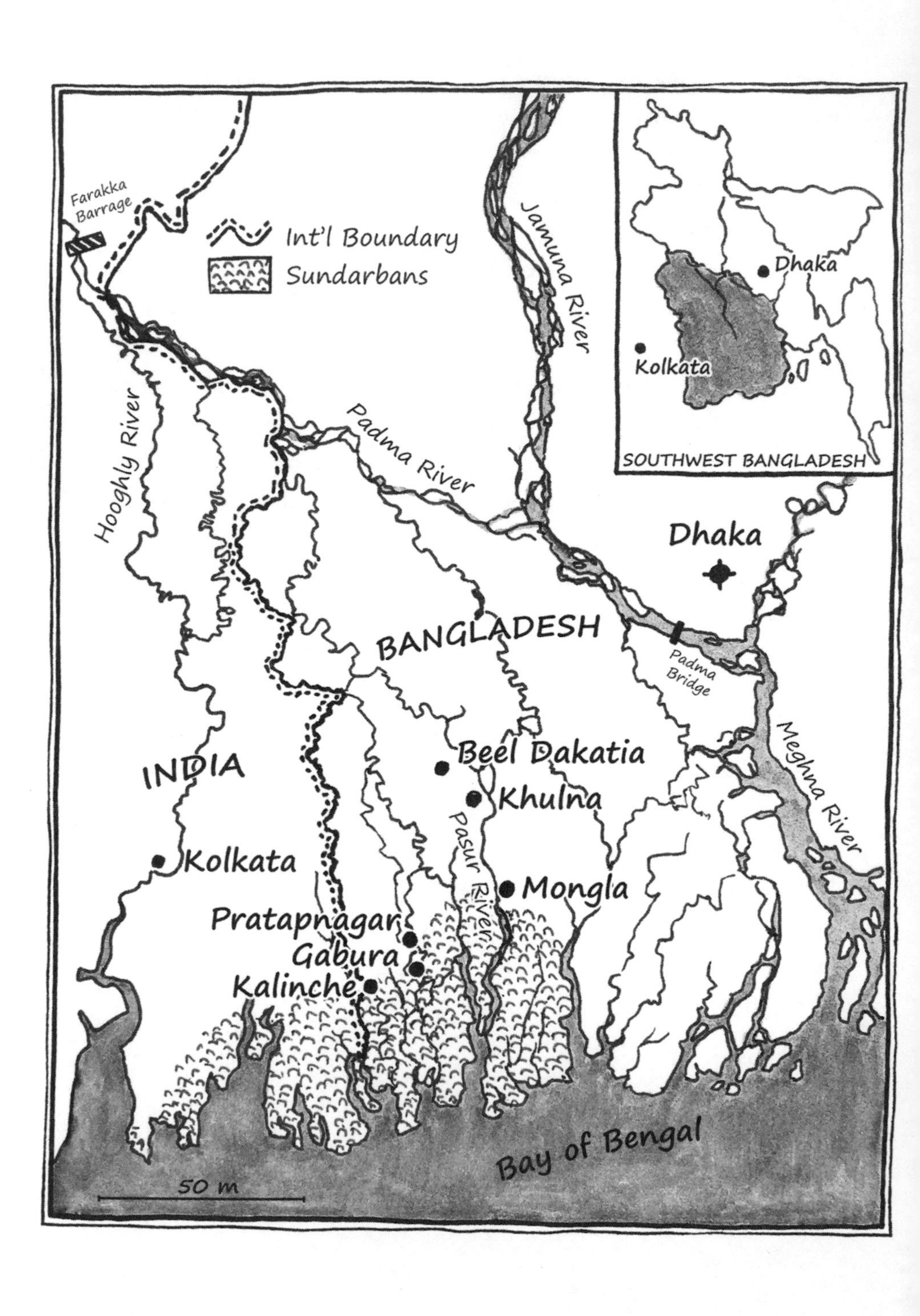

Farakka Barrage
Int'l Boundary
Sundarbans
Jamuna River
Dhaka
Kolkata
SOUTHWEST BANGLADESH
Hooghly River
Padma River
Dhaka
BANGLADESH
Padma Bridge
Meghna River
Beel Dakatia
INDIA
Khulna
Pasur River
Kolkata
Mongla
Pratapnagar
Gabura
Kalinche
Bay of Bengal
50 m

Part II

Pagal, by Any Other Name

Southwest Bangladesh

"When the winds shift, the fear comes," says Abdul Rahaman Gazi, who lives on Gabura, a small island at the mouth of the Ganges Delta. In Bangladesh, this shift happens twice a year, in March and September, when warm air blowing in off the bay brings fresh cyclone seasons. Gabura is protected from the torrents of tidal rivers by embankments modeled after a Dutch strategy for flood control, but these often fail. When they do, brackish delta water rushes in and lingers for months or years behind the levees. Gazi's home is perched precariously atop a crumbling earthen embankment, and he has watched the rising high tide cut into this feeble defense for years. "We're in trouble," he tells me; "our house is on the river."

"There aren't any beginnings," Burton said. "Nor any ends. It seems to me that man has engaged in a blind and fearful struggle out of a past he can't remember, into a future he can't foresee or understand. And man has met and defeated every obstacle, every enemy except one. He cannot win over himself."

—John Steinbeck, *In Dubious Battle*[1]

Word reached Karunamoyee Sardar that the men were approaching sometime around 10 in the morning. She immediately headed for the river. It was November, and dew was just lifting from the blades of grass shooting out of the gray Bengali mud. As she rallied the other villagers, her grandson ran at her knees, begging to come along. The child's mother caught his arm and pulled him back. Karuna strode west and climbed the embankment that held back the Habrakhali River. She dug her toes into its slippery earthen slope and pressed on to meet the incoming boats. She expected a confrontation.

Karuna was slight, with black hair and small eyes. Like other Hindu women, she draped herself in a brightly colored cotton *saree* and placed a small red *bindi*, or dot, made from mixing turmeric powder with lime juice, between her eyebrows. She was the mother of two sons and a daughter, and despite her stature, all agreed she was a force to be reckoned with. She was also educated, as much as that was possible in rural Bangladesh in 1990, and had remained active in the local *samiti*, which advocated on behalf of the rights of peasants.

Her family lived in small square house with woven straw walls and thatched palm roof in a village called Bigordana. It sat 60 miles north of the Bay of Bengal on a dagger-shaped island that came to a sharp point where the Habrakhali met the Badurgachi River. Some 10,000 people lived in communities like hers spread across the island, and they survived on what they could grow in the soil between these turbulent waters.[2] Floods were common, and so they built their homes atop elevated mud platforms and put their faith in an 11-mile-long earthen embankment that encircled the island. It provided a thin line of defense from the voracious bay tide.

Life on the edge of the world's largest river delta had never been easy. Karuna lived with drought, flood, hurricane, and crop failure by saltwater poisoning. Still, it was a stable if not luxurious livelihood—that is, until outsiders showed up with a simple fix that threatened to ruin all she had built.

In the 1980s, the Bangladeshi government and foreign aid organizations began aggressively promoting shrimp aquaculture as an adaptation to the problem of saltwater intrusion in rice paddy. Shrimp could be raised in ponds of brackish water that killed other crops, and unlike rice, most of which sold locally, it fetched a handsome price from Western buyers. Over the previous decade, as saltwater from tidal rivers had seeped into farmers' fields, green paddies on islands surrounding Bigordana had faded to pools of tepid, silver brine. Farmers were going all in on this new money maker, but Karuna was unconvinced.

From the beginning, the shrimp industry had been dominated by a wealthy few, most of whom lived hours away in Dhaka, the megacity capital that was home to 7 million people at the time. These absentee landlords didn't care that seeping brackish water destroyed the cropland around their shrimp ponds or that it poisoned shallow wells that supplied village drinking water. They even cut holes in the protective embankments to let saltwater in, which left everyone on the island defenseless against floods. Locals had little power to resist what they called the shrimp mafia that controlled shrimp processing, set its price, and sent its goons to wrest land from struggling rice growers when their shrimp ponds turned toxic with algae and bacteria, as they inevitably did.[3]

One of its bosses was Wazed Ali Biswas.[4] Hell-bent on expanding his shrimp operation into Bigordana, Biswas had been sending his goons to the island for months leading to November 1990, but their requests were preceded by word of the shrimp industry's harms. When his men came asking to buy Karuna's family's land, her husband rejected their paltry offer. When they returned, offering to build the family a new house and then again with promises to find her husband a job at a brick factory, he again refused to move. Biswas's men grew furious. Karuna took to organizing the community.

On the morning of November 7, five trawlers approached the island carrying Biswas's strongmen on their way to cut the embankment, flood the land, and leave its residents with little choice. Karuna went to meet them. As she headed southeast toward the island's lowest point, wide rice paddies and clusters of homes elevated on bamboo supports spread out on her left. On her right ran the brown and unceasing Habrakhali River. Other villagers fell in behind her, and together they formed a phalanx marching with fists raised through a tunnel of green rain trees to where the invading boats had landed.

The groups approached each other apprehensively. Biswas's agents were not wealthy elites but men finding their own desperate way through the delta's poverty; Karuna hoped a show of forceful solidarity would send them retreating across the river. The villagers stomped their sandals in the mud and chanted slogans in support of their right to the land, but the men had come armed with guns, knives, and homemade bombs, and the morning took a horrific turn.

Shots pierced the morning stillness. Bombs clattered along the rough brick road and erupted in fury. From her position at the front, Karuna was swept up in the mayhem. An explosion at her feet seared the air where she stood and spit shrapnel into roadside trees. When she fell, the other villagers splintered in panic. Two of the strongmen ran forward into the clamor; they gathered her lifeless body and dragged it back to their boats. Then they gunned their engines, carving wakes in the milk chocolate river as they sped downstream toward the Sundarbans. In this impenetrable and unpoliced forest of coastal mangroves, they would hide out amid outlaws and Bengal tigers.

The dust and the acrid scent of gunpowder had barely settled when the police arrived. They questioned a few of Biswas's men who had stayed behind to kick dirt over the gore remaining on the roadway. They told the police they had killed a dog—miserable animals that get little sympathy in Bangladesh—but above them, long black hair blew in the breeze from a piece of skull clinging to the branch of a rain tree. The police noted this and walked away.

Despite the crowd of witnesses, forty of whom suffered injuries in the skirmish, fourteen years passed before any of Biswas's men were brought to trial. In 2007, a court in Khulna sentenced twelve of them to life imprisonment and issued each a fine. By 2008, all twelve had been released. Meanwhile, Biswas was elected to Bangladesh's parliament in 1994 and died of a heart attack at his home in 2003, after being honored with the National Export Trophy and the President's Medal for his contributions to agriculture and social services.[5] Karuna's family never received her body, so she could not be granted a proper Hindu cremation. Locals assume she was dumped in the river.

~

I arrive in Bigordana in March 2022 and hear this story from those who lived it. Karuna's daughter and daughter-in-law set a plastic chair for me beside their garden. Afterward, I tour the village with a local farmer and find a land that looks, smells, sounds, and feels richer than those I passed through to get here. The air is thicker, sweeter, and more pungent. It buzzes with the calls of Oriental magpie-robins, black and blue flycatchers that perch in the lush palms and rain trees that crowd the sky. I wander past homes with open doors and follow the road to the outer fields where birds swoop and dive over vibrant fields of green paddy. Elsewhere in southwest Bangladesh, lands are gray and almost treeless, because in the years since Karuna's death much of the region has been flooded and entombed in the brackish shrimp ponds she died resisting.

Shrimp is now the country's second most valuable export commodity, after garments, and some 35 percent of it ends up in the United States.[6] The industry has brought wealth to landowners like Biswas but has left small-scale farmers even more vulnerable than before. Having missed out on the windfall, they've been saddled with all of the painful side effects Karuna had feared. Three quarters of those living in southwest Bangladesh drink water that's dangerously saline. It causes rashes, liver disease, and diarrhea. Even the cows, which are sacred to Hindus, look mangy and thin.

But not in Bigordana. On this clear and cool day, I see how well the island has fared since her sacrifice. Fearing prosecution for her death, Biswas opted not to press the issue, and the embankment was never cut. Locals consider Karuna a martyr and built a memorial at the spot on the embankment road where she fell. There's a colorful mosaic that depicts her marching tall at the head of the procession, black hair down and fist raised. Each November 7, they gather there to remember her.

At first, I file Karuna's story away with others about commercial aquaculture and food sovereignty, sea-level rise, and climate change. But the more I see of Bangladesh, the more I realize it exposes a deeper wound in the local history, one that traces the heavy-handed involvement of foreigners who sought to wring all they could from this rich land by molding its clay into whatever form suited their interests. It's a story about corruption on a grand scale and an environment unlike anywhere else on Earth, with lessons that echo into our time. And if there's a way to summarize it all in a single word, it's with one that was loaned to Bangla from Europe, like so many of the country's troubles, because when people talk about what happened to Karuna they don't talk about Bigordana or any of the island's other villages—they talk about Polder 22.

Polder is an old Dutch word for "a land covered in puddles, wetland, marshy ground." Today, it denotes land that has been protected—some prefer reclaimed—from water through the construction of a raised embankment, or levee, and it has now become so intertwined with life in the Ganges Delta that it might as well be Bangla.[7] A polder is a small, flat island encircled by walls that rise anywhere from a few feet to well overhead and slope into surrounding waters. About a third of the Netherlands lies below sea level, and so the Dutch developed polders to claim land from the frigid North Sea.[8] Later, their strategy was put to great use in Bangladesh, but the engineers who did overlooked a key difference between the countries: In Bangladesh, 170 million people share a space slightly smaller than Illinois with the largest river delta in the world. It consists of more than 230 shifting rivers, which together carry enough sediment to keep the land above sea level, if permitted. This is not so much a world pestered by water as one defined by it.

The people who have lived beside the Bay of Bengal have dealt with its reality longer than Bangladesh has been a country, longer than it was a colony of England or a Muslim fiefdom. Through it all, they found ways of coexisting with water, but the polders represent something new: an effort to contain and control. I happen to be traveling through Bangladesh exactly fifty years after its bloody war for independence. In every city I visit, in every village square I linger, loudspeakers blare the voice of the father of the nation, Sheikh Mujibur Rahman, often called *Bangabandhu*, or "friend of Bengal." In March 1971, he gave an eternal address to a newly free people at a time when their country was derided as the "basket case" of Asia. Much has changed since then.

In 2022, per capita gross domestic product surpassed those of both India and Pakistan. Quality of life has seen a meteoric rise, thanks especially to social programs focused on the needs and care of Bangladeshi women. Farmers here produce more rice than all but three other countries.[9] Parts of Dhaka shimmer like Tokyo and New York. The country appears on the cusp of joining the ranks of the developed nations, but a problem it did not cause and cannot fix threatens its progress.

Bangladesh is notoriously vulnerable to the effects of climate change: sea-level rise of as much as 5 feet in some places predicted by 2100, shifting rain patterns, drought, monsoonal floods, cyclones, disease, crop losses.[10] These torments cost the country about $1 billion annually and send millions of people fleeing for safety or opportunity, often in Dhaka, where housing, jobs, and basic services are stretched thin.[11] Over the coming decades, the World Bank estimates that Bangladesh will need $5.7 billion to adapt to these threats.[12] In the media and in pleas for aid, these problems are often presented as the novel and direct effects of climate change. But that obscures the full truth. When European voyagers arrived in Bengal in the sixteenth century, they encountered a land whose agricultural riches rivaled those of the Nile Delta. Within three centuries, Bengal descended into a famine-plagued charity case, and much of the blame can be attributed to the unintended consequence of a past adaptation.

About half of the country's southern coastal zone—an area larger than the state of Maryland—has been cordoned into 139 polders that contain more than 8 million people.[13] These fortified islands were supposed to insulate fishing and farming communities like Bigordana from the waters surrounding them; in fact, they have added to the troubles.

"Trying to find land for 170 million people is an enormous problem," Dilip Datta told me when I visited his office at Khulna University. "The assumption around the Bengal Delta is that we are fixing these issues with infrastructure. In truth, our actions are responsible for perpetuating this problem." Datta began as a geologist and gradually worked his way through museology, geochemistry, and biochemistry before focusing on issues affecting rural women. This led him to food security, soil science, and ecology. In the end, it all came down to the hydrology of the Ganges Delta and the dubious means that have been used to control it. "We want to reclaim land because of population density. We need more, more, more. The people still suffer," he said.

For centuries, new people from new places have taken the same approach to fixing what they thought was wrong with this place. The cycle has resulted in failure for the same reasons. "The embankments provide only pseudo security. They work against our best interest," Datta told me. Floods that were once common but manageable have become less frequent and more devastating.[14] The fabric of daily life and social structure have experienced traumatic upheaval—Karuna's story and the countrywide conflict over brackish shrimp farming is just one example—and perhaps most insidiously, efforts at long-term adaptation to climate change have been undermined. All this attests to a glaring mismatch between a solution and its problem, but as adaptation efforts ramp up, governments and foreign aid organizations continue to follow a path set centuries ago.

"It's the definition of—what's the word? In Bangla we say *pagal*," Datta said as his hands and eyes probed his office for the word. "Madness!" he said at last. "It is the definition of madness to do the same thing over and over and expect a different result."

Like the people they were made to protect and the rivers they were meant to contain, the embankments are complex, rife with turmoil, and—

in opposition to their designers' intent—ever-changing. To understand why they're here and how they've affected the country, I travel from Dhaka through the coastal southwest, where the long-term effects of this infrastructure and the more recent impacts of climate change are apparent. Although Bangladesh is geographically small, its lacing rivers and shifting mud make traversing it a practice in versatility and patience. With adept local guides, I travel by plane, car, bus, motorcycle, boat, and sandaled foot, always trying to stay dry and always getting wet.

The Polders of Pratapnagar

Bengal spreads across the eastern part of the Indian subcontinent, encompassing modern-day Bangladesh and a piece of eastern India. Its politics have navigated the historical gamut, but the land has always been dominated by two rivers: the Ganges and the Brahmaputra. Together, they drain the Himalaya Mountains, which began forming 50 million years ago when islands carrying India, Bangladesh, and Sri Lanka drifted from the coast of Australia and collided with Eurasia. The planet has been watching that train wreck in slow motion ever since. As the two landmasses have crumpled and fissured against each other, Mount Everest has risen above everything else on Earth, and the Bangladeshi delta has sunken a bit while sloping toward the sea.

The Ganges is the mightiest river.[15] It spills from glaciers on the Himalaya's southern slopes and flows east across the Indo plain, collecting tributaries and the sins of millions of Hindus along the way. Crossing into modern Bangladesh in the country's northwest, it takes on the name Padma, given by locals who see the river as the "destroyer of creation." The Brahmaputra is the son of Brahma, the Creator. Its journey begins just north of the Ganges on the other side of the Himalayan ridge, then parallels the range moving eastward across the Tibetan

Plateau before hooking around the flanks of the mountains and dropping into Bangladesh at the country's northern border. As it braids wildly across prime agricultural plains, it takes the name Jamuna, or Holy River. The two rivers come together about 30 miles west of Dhaka and are soon joined by yet another huge waterway, the Meghna, named for the clouds that drape across the high eastern hills where it originates. It operates as the delta's vena cava, collecting the others before spilling into the Bay of Bengal. The whole system drains 419,300 square miles, just 7 percent of which is within modern Bangladesh, and averages a discharge of 1,086,500 cubic feet per second, with flows peaking during the summer monsoon.

But that's putting it simply. The delta is made from silt, sand, and bits of the craggy Himalayan cliffs that have washed down and coalesced into deltaic mud that breaks apart easily, like garden soil sprayed with a hose. As this sediment piles up, it blocks old channels and sends water draining in new courses. Dissolved minerals from the mountainous debris load the delta with nutrient-rich silt, and when it floods it spreads this natural fertilizer across the vast sedimentary plains. Combined with nearly constant sun and a long and humid growing season, the Ganges, like the Mississippi, is home to some of the richest soil around and yet, at the same time, some of the poorest people.

Near the Bay of Bengal, winding rivers have carved the soft Bengali mud into a smattering of elongated islands and divorced islets. New lands appear where the rivers dump them—they are called *chars*—and old banks crumble away just as quickly. This is a fact of life on the delta, as are bad roads. The first time I visited Bangladesh, I asked a companion where we could find an embankment. We were walking along in the country and they pointed to our feet. Because they are often all that remains above water during floods, embankments double as roads that encircle and bisect the polders they contain. On big rivers near large cities they might be robust and concrete, but most of the time they're unremarkable sloping mud walls. I had been traveling along them all day without realizing it.

Bangladesh is far more complex than foreign media would have it appear, but it is rife with troubles. The challenge is disentangling interwoven causes—climate change, colonialism, corruption, industrialization, infrastructure, bad weather, geology—to find their sources. Today, I'm about eight hours southeast of Dhaka searching for signs of the polder's impact with Riton Camille Quiah. A Bangladeshi from a southern town, he carries two pairs of Crocs—one for the mud and one for the city—and never passes up tea and a cigarette. The tide is low when a small ferry arrives to take us across the Kholpetua River. There's a 25-yard stretch of wet, sticky mud between the dock and where the boat waits, so we roll our red Honda Hero motorcycle down a string of slippery wooden boards to a narrow gangplank where it teeters several feet over murky water. The boatman jumps into a hole in the center of the wooden *dingi* and cranks the flywheel on a leaky diesel engine that sputters, spits out a plume of thick, black smoke, and roars to a deafening clamor.

On the other side, Quiah and I climb on toward Pratapnagar, which is located in a polder that's defined on its east and west sides by two north–south snaking tidal rivers and made into an island by smaller crossing channels. The embankment road is sprinkled with shattered red bricks that toss the bike so that a day of riding beats my ass to a pulp. Off my right, within the embankment, the land is a flat expanse of tan mud with cracks wide enough to fit a hand. A few tree stumps poke up from the middle of bone-dry fields. They are all that remains of palm stands thrashed to bits by back-to-back cyclones that left this land under several feet of brackish water for most of the past eighteen months. Now it's a parched expanse of fractured mud.

In the 1960s, Western engineers, foreign governments, and aid organizations flocked to the Ganges Delta to save its starving people from monsoon floods and cyclone storm surges. The crown jewel of that effort was called the Coastal Embankment Project, under which the US Agency for International Development (USAID) spent $278 million building the first 103 polders behind nearly 1,000 miles of embankments. Quiah and I are bouncing along on a Chinese-made motorcycle through the results of their work.

The sun beats down on us as we cross wide expanses of crusty fields. Somewhere near the middle of Pratapnagar, we dismount and walk a dirt road past store shacks with tin roofs and concrete walls pocked by sea and rain. A few lonely palms stand motionless above us in the stale heat, and a stray black goat wanders, baying for its trip. It's been at least an hour since we've had tea, so we dip into a *cha*, or tea, stand. Quiah lights a cigarette, and we order a plate of fried *shingara*. Two boys walking home from school notice us and poke their heads in to see what we're about. Both sport neatly trimmed black hair and collared blue school uniforms. The eldest, Rakibul Hasan, is shyly soft-spoken but brave enough to try out his school-learned English on a foreigner. He tells me that he was twelve years old when Super-Cyclone Amphan came here on May 20, 2020. He remembers it well. "I was very afraid. My family too. We wondered whether we would survive," he recalls.

Perhaps I shouldn't be surprised to learn that people who have long occupied a land plagued by cyclones don't scurry in fear at their approach. If the weather report predicts landfall at seven in the evening, farmers will tend to their fields until five. Officials have been encouraging people to seek shelter earlier, but this goes against the culture. However, Hasan's family saw that Amphan was a terribly different storm and organized quickly. They grabbed what they could from their small yard and brought it inside the house, hoping the thatched walls would at least abate the wind. They were huddled on the communal bed when the water came lapping at their kitchen stove. In rural Bangladeshi homes, beds are usually square wooden platforms elevated a couple feet off the dirt floor and padded with a thin blanket. During floods, it's just about the only place that remains out of the sink.

Outside, all the world was water as Amphan melded the sea and air. It started with a light rain that pattered the flat leaves clinging to the family's trees. Then fat drops kicked up dust in the yard. Soon, the sky went white and Hasan's ears were filled with a strange whirring sound like an approaching army of oversized bees. By then, even temerarious farmers were running home, gripping their soaked lungis closely. Amphan was the most powerful storm ever recorded in the Bay of Bengal. Its winds reached 170 miles per hour and waves crested 16 feet.[16] Its swirling outer arms lashed at Hasan's walls and roof in unrelenting surges. The boy prayed quietly, clinging to the thin fabric of the bed. Overhead, palm trees bent sideways until they were torn out of the earth altogether.

As Hasan tells his tale, men crowd around us at the cha stand, each with something to add. They explain that the drama did not end with the cyclone; what came after was far worse. Amphan's immense power pushed the Bay of Bengal's surging waters inland, and the rivers encasing Pratapnagar quickly rose to flood stage. On the east side, the Kopothakho River beat violently against the island's mud embankment at a bend where years of erosion had weakened it. When the wall finally gave way, seawater spilled into the low polder through a gap 500 feet wide, wiping homes, trees, and roads clean off the map. A narrow canal bifurcating the island swelled into a raging torrent that carried off the only bridge connecting its halves and then continued to spread across the land. Three men jumped into a wooden dinghi to go repair the broken embankment, but the turning tide sucked water back toward the sea faster than it had poured in. Caught in a whirlpool, they gripped the rough gunwales as their boat spun and lurched until at last it was swamped and they were all taken under.

Under the relentless heaving of heavy storm tides, the island's embankments failed in six places, totaling 30 miles of breaches. Families like Hasan's were flooded out all across the island, clinging to corrugated steel roofs and floating on debris while the tide came and went at will.

"I didn't see my friends for several days after. I could only leave my house at low tide," Hasan tells me. When the waters receded, he emerged into a world that looked nothing like the one he remembered. The homes all around his were gone. The island's thick palm stands had vanished, save for some stripped trunks whose bareness only added to the sense of isolation. Whether because his home was sturdy enough or he had so little to lose, Hasan laments only the loss of his cricket stump, a prized possession but an easy one to replace.

Hasan looks tired of the limelight, so I thank him and set out to see the damage for myself. I try to imagine green leaves swaying over rich paddy, but all I see are dead palms rising like bent cigars with rough-cut ends. Rain trees have been stripped and cracked. A few mangoes and coconuts stand dead. On them, the high-water mark shows clearly as a white line at my shoulder. Saltwater inundation stopped farming and tainted groundwater supplies. Schools were shuttered, and people went packing. Locals managed to repair the eastern embankment just before May 2021. Then Cyclone Yaas breached the wall on the west bank. Marooned again, they survived on government food rations and disaster relief dollars while they waited for the waters to subside.

The government of Bangladesh's response to calamities like this has been to seek more money for more and bigger embankments. Foreign nations and investors are quick to answer the call—for reasons that I will soon discover—but when I visit Pratapnagar, I don't see the grand success story of flood prevention. In the years since the Coastal Embankment Project, flooding and damages have actually increased in the coastal zone.[17] In 1953, flooding ruined half a million tons of crops. In 2007, the damage was more than twice that. This is partly a result of early planning that designed embankments for twenty-five- to fifty-year flood events when the last six major floods (1987, 1988, 1998, 2004, 2007, and 2017) have all fit the characterization of hundred-year events. Also, increasing damages can be blamed on the sense of confidence that the walls have engendered. Just as I witnessed in Japan, the embankments have made farmland materialize where there was once a river, and their presence has attracted more people and capital to flood-prone areas, thus raising the stakes when the walls fail. During the 2017 flood, for instance, areas along the Jamuna River with higher levels of protection experienced higher rates of mortality.[18]

It's not that the embankments have failed entirely—just the hint of this makes many of my academic sources get nervous—but the numbers show that the polders have not necessarily decreased flooding. In 1988, 1998, and 2004, 17 percent, 24 percent, and 27.7 percent of the country's embankments breached, respectively. Researchers have calculated that the extent of inundation from Cyclone Sidr's storm surge in 2007 was 17 percent greater because of the embankments. "We conclude that whilst polders have provided protection against storm surges and fluvio-tidal events of moderate severity, they have exacerbated more frequent pluvial flooding and promoted potential flooding impacts during the most extreme storm surges," writes Mohammed Sarfaraz Gani Adnan, of the University of Oxford.[19]

Meanwhile, the claim that polders are the best means of alleviating poverty is untethered from reality: The share of income held by the poorest 40 percent of Bangladeshis declined from 17 percent in 1991 to 13 percent in 2016, a period in which, unsurprisingly, the wealthiest 10 percent saw their share increase.

~

The polder strategy has provided the rural poor with one thing: land, but not always in the way it was intended. An estimated two million people now squat on riverside government-owned land, often directly behind or on top of the embankments. Their situations are beyond precarious. Leaving Pratapnagar's center and approaching the Kholpetua River, I climb up the slippery slope of an embankment where I find Muhammed Habibul Rahaman. He lives with nine family members in two small shacks just out of reach of the water during the dry season, and his mother-in-law emerges when she hears a stranger's voice. She stares into me from behind a dark green *sari* held across half her face.

"Five years ago, our house was over there," she tells me, pointing toward the middle of the slate gray river. It's low tide, and I search the mud flat to make out the remains of some trees and what looks like a wooden utility pole where it shouldn't be. The family has already changed locations four times, always moving inland to stay out of the river, and now they must move again. Seventy other families have already resettled inland because they were living in the way of a new embankment going up at the river's latest edge. Their house is pieced together from scraps of corrugated sheeting and bamboo poles. Debris, bags, fishing nets, clothing, tree bark, and nepa fronds hold down the roof. The whole structure is anchored into the mud embankment by several thick ropes strapped across it. They invite me inside, and I stoop through a narrow doorway into a space barely large enough for a bed and some shelves.

Large rubber buoys, blue plastic jugs, and bulging white aid relief sacks are stacked all around. Clothing hangs drying from every available bamboo strut. From the far corner, two black eyes stop me dead in my tracks. A young woman in a crimson sash sits cross-legged on the bed. Her exposed black hair is pulled into a bun, and she holds an eighteen-month-old baby in her lap. Her piercing, concerned stare quickly softens as I step backward, and she angles the child so I can see his dozing face. I take a quick look around and retreat toward the light, feeling like I've intruded. Outside, Rahaman says he has managed to squirrel away about US$5,700 for new land inland. They will carry everything they own on their heads.

I follow the embankment farther along the river's edge as it undulates over high intact sections and low crumbled gaps. Along the way, I encounter three men working on repairs. By now the afternoon heat is oppressive, so the men agree to a breather and a quick conversation. Abu Zaffor is sixty-three, wiry, and deeply tanned. He works barefoot, wearing a weathered lungi and a white goatee. He and the others have wrapped their heads with something like a flat turban that secures a sturdy circular platform for carrying heavy loads. They form a train: One digs mud from inside the embankment and loads it into a woven reed basket. Another hoists the basket onto his head, walks it a few strides up the slope, and passes it off to Zaffor, who carries it to the wall's forming edge, dumps it, and returns for more.

Zaffor says 2,000 people have been working to fix this wall since the waterline finally withdrew in November. They're all local, pulling six-hour shifts beginning around dawn. A contractor pays them US$4 a day. Zaffor has lived his entire life at Pratapnagar and says he's seen the embankment move six times already. Before Amphan, it stood about 35 yards into the river. He shields his eyes from the sun's glare and points that way with a thin arm. "By the grace of Allah, this new embankment will last a long time."

I can't imagine what would give him that idea.

Everyone I talk to recounts repeated upheavals in a place where the only constant appears to be change. Time and again, each has packed what they could carry on their heads and moved inland with the river lapping at their heels. On the delta, everything moves. This fact has been accepted, if not necessarily embraced, by rural Bangladeshis, but it doesn't appear to have penetrated the thinking of outsiders.

As I walk around Pratapnagar, I keep coming across dilapidated brick buildings. They stand out against all the palm shacks, and they grab my attention because each is empty, hollowed out, roofless, and slumping. They might have been handsome structures at one time, but now they look like the kinds of places my childhood friends would gather to smoke pilfered cigarettes and peruse outdated porno mags. Some foreign aid group built them, locals tell me, but they didn't consider how the mortar would dissolve in the coastal sea air. Why build something so ill-suited to the region? Why invest so heavily in permanence on a land that demands flexibility?

I leave Zaffor to his work, hauling mud that came from rock, which was thrust upward from the earth's mantle, then worn to sludge and washed toward the ocean, where it gathered and piled until it rose above the water, becoming land that is collected in a woven basket and packed by hand into a wall that stands in the shadows of six more just like it, all gone now. So much for stability.

In town, a young woman stands in her doorway and watches me cautiously. She is Aklima Khatun, and although she looks much older to me, she is only thirty-five. She and her family had lived in a house near the river until Amphan washed it away. Many of her neighbors have left, but Khatun remains because she has no other option. "My whole life has been water," she tells me, but "this was the most damage we've seen in our lives." As the water rose, she watched the embankment fail. "I have seen river erosion my whole life. The embankment has broken many times, but the last time was the worst." Afterward, she moved to this abandoned, open-air fish market. She and a few other desperate families tried to fill in its walls with palm fronds and wooden boards.

She leans against the wooden doorframe and searches me with tired brown eyes through an opening in her burgundy sari. Behind her, I can see the bare right foot of another woman who remains hidden but listens closely to our conversation. In the dirt outside, Khatun's young daughter fries red chilies over a mud stove. Nodding to the small girl feeding sticks into a flame, Khatun says, "Her brothers are working for all of us. One works at a cement plant, another in brick factory. The family is all spread out now." Her husband drives a rickshaw someplace where people have money to travel. Before the storm, she had kept ducks and cows. "When I had my own house I could feed my children. Now I can't because we don't have anything." I don't get the sense that she has any plans to rebuild her home. She doesn't seem to have much hope at all.

It used to be that cyclones came once every few years, and we had time to rebuild, she says. "Now almost every year there is a cyclone coming." She knows another storm will pound Pratapnagar in the months ahead, but what can she do to prepare? And what is the point? I ask who or what she blames for her apparent misery. "*Pāni*," she says. "Water has been the reason for my hard life."

Toward the end of the day, Quiah and I find a sturdy concrete building that serves as a cyclone shelter and seat of the local government. We climb to the second-story office of Pratapnagar's chairman, Hazid Abdul Dowd, who greets us cordially from behind a wooden desk. He wears a clean white tunic and white hat and is framed by a window facing the parched island. The room is filled with other men, official aides and modestly dressed locals who sit in several rows of folding chairs. Each has come to ask for help or to air a grievance, but Dowd generously agrees to let me cut in with a few questions, and they all listen intently.

He tells me that Pratapnagar's embankments were in disrepair before Amphan. "You can be walking along and it will crumble," he says. When I ask why they hadn't been repaired, he is exasperated. They had repeatedly petitioned the government's Water Board for help, but none came. Picking up on the chairman's frustration, the other men chime in with complaints of their own: Their population is 40,000 and they have only enough room for 6,000 people in their cyclone shelters. The soil is too salty for growing rice, and the waters don't produce enough fish. Their only sluice gate hasn't worked since 2009. There are still 28 miles of embankments that are unsafe.

"Five hundred families have left the island."

"There are no jobs for local people."

"They are not making a strong embankment in the right place."

"The money for the embankment is not going where it needs to go."

"From top to bottom, everywhere is corruption."

Corruption in Bangladesh is like the weather: It might not blow you away, but it's bound to get you wet. In the usual way, contractors ingratiate themselves with politicians to secure lucrative infrastructure projects, then make a few extra bucks by trimming the plan and pocketing the unspent budget. A wall doesn't end up being as tall as promised. Nothing is done to improve an embankment's sandy foundation. No one returns for maintenance. This type of corruption is by no means specific to Bangladesh—some reports estimate that it will cost the world as much as $5 trillion annually by 2030—but Bangladesh confronts disasters with a documented complacency.[20]

So well habituated to catastrophe, it utterly neglects the long-term challenges, found Transparency International Bangladesh, a locally operated anticorruption organization.[21] Six months after Cyclone Amphan, for instance, the Water Board had yet to repair at least fifty-four points of failure along a 144-mile stretch of damaged embankments in a district with a population of 250,000 people. As a result, 20,000 were still homeless. Even cyclone shelters—the last resort for millions in their worst moments—fall prey to corrupt engineers who construct the buildings near their own homes and beyond reach of fishing village residents who need them. In Koyra Upazila, across the Kopothakho River from Pratapnagar, Transparency International Bangladesh found that three quarters of one project's budget had been embezzled, and the work was never finished. A few miles south in the impossibly vulnerable community of Gabura, a contractor was paid to raise an embankment by 3 feet but pocketed the cash without completing the task according to the work order, and the whole island flooded during Amphan. In this case, locals accused the contractor and Water Board officials of embezzlement.

I mention a large rehabilitation plan that I've been hearing about and ask whether Pratapnagar is benefiting from it. "We have been hearing these types of things for several years, but until we see it with our own eyes, we don't have much faith," Dowd says. "We are seeing nothing happening."

I stare across a room full of eyes, each intent on me, a stranger who dropped into this world with nothing to offer but questions. At a loss, I ask whether any of the men knows what happened before the embankments were built. One, sixty-three-year-old Abdul Salam Morel, responds: "Before the embankment, my grandfather made a plot by piling mud by hand. It was smaller. The area was famous for rice farming then. There was no shrimp farming. My grandfather used to rebuild the embankment every year. We had rice and fish in the paddies. In my childhood the river was forty to fifty hands deep; now it is only ten hands to the riverbed and the water level is higher. The rivers are filling up because of siltation. Nowadays it is a dying river."

The Curious Death of Rivers

Levees aren't new technology; surely by now we've figured out how to pile mud on riverbanks. So why have the embankments failed? Corruption is a large part of it, as is lack of maintenance. Government officials and representatives of some international aid organizations often place the blame squarely on climate change, saying that sea-level rise and cyclones have outmatched their walls. This is only part of the story. The rest lies with the phrase *dying river*. It refers to a phenomenon specific to southwest Bangladesh, something, like mortar in sea air, that was known by some and overlooked by others.

A British engineer named Sir William Willcocks came so close and yet so astoundingly far from deciphering the mystery; he deserves credit, at least, for raising the question. A civil engineer at the English Colonial Empire's peak, he was born in a tent on an irrigation canal in India in 1852 and spent his life building some of the most advanced irrigation works of the time. He blamed England for making a mess of things when it ruled the Indian subcontinent, and he accused his countrymen of ignoring the ancient knowledge of the place.[22]

The dead rivers are coastal rivers, like the Kopothakho, found in eastern India and southwest Bangladesh. For some reason, they lack an upstream connection with the Ganges flowing out of the Himalaya, and so, rather than draining mountain runoff, they ebb and flow tidally with the Bay of Bengal. Willcocks noticed that they were broad, shallow, and running in nearly parallel paths to the sea. They were also somewhat evenly spaced at "the right distance from one another for purposes of irrigation" by overflow flood. And so he reasoned that they were actually an ingenious network of canals, cut 3,000 years ago under the direction of some ancient Bengali king who must have learned the method from travelers out of Mesopotamia. Just before he died, in 1932, Willcocks published a series of lectures in which he argued that the railroads and embankments that the British had built across Bengal had thrown a wrench in this native means of irrigation.

Departing Pratapnagar, I travel to the industrializing port town of Mongla. It's a popular gateway to the Sundarbans for the few tourists who venture there, and I'm looking to take a ride on a dying river. In the morning, I wake to the *adhan*, the Muslim call to prayer, which echoes from minarets in neighborhoods across the city. I roll out from under my mosquito net and head for the dock to board a gleaming white tourist boat, called the *Kokilmoni*. When I arrive, it's already packed with a cadre of college students, hydrologists, and geologists from Dhaka and Columbia universities. The boat has an unused berth, and so I tag along as the scientists work their way down the Pasur River, mapping freshwater aquifers under the delta, because, apparently, in this thoroughly drenched landscape, millions of people can't get enough to drink.

Onboard, I seek out Michael Steckler. A polymath from Columbia University, he sums up his career by saying that he came to Bangladesh twenty years ago to study sedimentation and has returned over and over again to explore rabbit holes he discovered on each visit. Below deck in the *Kokilmoni*'s galley, he sweeps aside cha cups and unrolls several large maps across a table. They are false color: flat blue for water, neon green for vegetation, and wine red where trees have been cleared for farms and development. The crew uses them to find ground suitable for deploying their instruments. For me, the braided channels, which are juxtaposed in such stark hues, jump out as an azure venous system splicing an intricate puzzle of green islands.

The Pasur River maintains a tenuous link to the Padma River. In time, though, it could go the way of its western sisters, which have had their heads cut off to become the dying rivers Willcocks failed to explain. What satellite imagery, riverbed mapping, and more than a century of evolving earth science have taught us is that these waterways lack an upstream source not because they were human-made canals but because the Ganges has been crawling eastward and leaving a trail of amputated appendages in its wake.

A mere thousand years ago, the Ganges entered the bay through what is now a putrid sewage canal called the Adi Ganga on the outskirts of Calcutta. Back then, the now mighty Padma River was an insignificant tributary far to the east. Over centuries, though, silt choked the Adi Ganga upstream, and the Ganges shifted east to an easier outlet. This cycle repeated until the Ganges, always searching for a clear path to the sea, linked up with the Padma sometime around the eighteenth century.[23] As always, the key to this mystery is right beneath my feet: mud, or, more accurately, sediment. The delta's rivers carry 1.84 billion tons of the stuff to the bay each year.[24] Most of it is coarse silt, gray illitic clays, and sand-sized pieces of Himalayan granite that tumble and roll in turbid waters, giving the rivers their milky complexion. But there is much more to the hydrology of southwest Bangladesh than its surface reveals.

I had long accepted that the Ganges carries sediment washed from glacial cirques high in the Himalaya Mountains to the sea, and I've seen countless references to that sediment filling channels in southwest Bangladesh. But if these rivers are in fact cut off from the creeping Ganges, where does the sediment come from?

The Bay of Bengal, it turns out.

The Meghna River carries most of the system's sediment to the bay through its outlet in eastern Bangladesh. Outwash peaks with the summer monsoons at the same time that winds shift from cool northerly breezes to warm southerly gales, signaling the beginning of cyclone season. These summer landward winds whip the bay's surface current into a counterclockwise gyre, which carries some of the oceangoing sediment westward along the coastline. Floating on the tide, the sediment pulses back into the country's decapitated rivers every twelve hours.

The tide is key. Twice a day, it reaches deep into the country. At its high mark, its velocity falls to zero, and for a quiet moment it just hangs. In that brief interim between coming and going, the tumbling particles of clay and sand settle like undissolved sugar in a cup of tea. Over the eons before the embankments, rivers spilled their banks seasonally and spread their sediment across the deltaic plain. Then, when the tide turned, the ebbing waters scoured their riverbeds, clearing the way to the bay. But the embankments changed this relationship. By not allowing minor seasonal floods, they captured sediment within the channels, where it has piled up. The riverbeds have risen, and rather than dredging downward their waters have spread outward, eroding the banks at a vicious pace.[25]

As night falls on the *Kokilmoni*, the heat finally subsides. Steckler and I retreat to the top deck with a bottle of smuggled rum. Bangladesh is a strictly conservative Muslim country, and finding booze anywhere is a chore. Rural parts are painfully dry. The news that he has some makes me forget all about the biting clouds of mosquitoes. We anchor close enough to shore to hear engines revving and rickshaws honking in a waterside village. Giant flying foxes swoop out of the darkening forest to feast on swarms of insects hovering above the lethargic Pasur.

Steckler adds a bit of coconut water to his liquor and says that after using many approaches over many decades, he and a multitude of scientists from around the world are finally beginning to see the whole of the Ganges Delta. "It's like the blind monks examining an elephant," he explains. "Each system measures a different part of the story."

In 2017, an ecologist from Louisiana State University named Carol Wilson quantified profound changes since Bangladesh began building embankments along more than 3,100 miles of southwest coastal rivers the 1960s.[26] She found that about 375 river miles had been choked with sediment. This has disturbed fishing, diverted commercial shipping through ecologically sensitive but more navigable waterways in the Sundarbans, and collected precipitation on the landscape. Her work was key to understanding what was happening in the rivers.

If Steckler has groped around one part of the elephant in particular, it's the settling of the land. The *Kokilmoni* has brought him to parts of Bangladesh where no roads go, so he could deploy hundreds of global positioning stations to track subsidence through slight changes in elevation.[27] He and others have measured compaction from earth cores taken out of deep drinking water wells. They've unearthed Hindu temples, which were built above ground centuries ago but have since sunken below the jungle's surface. They've studied ancient salt kilns that prove the coastline was well beyond its present edge three centuries ago. They've used surface elevation tables, tide gauges, radar, satellite observations, and old maps to tell the story of a delta that is actively sinking—but not only because of climate change.

The entire Bengal Basin has been subsiding since the Eocene, more than 30 million years.[28] Subsidence happens here in three ways. At the surface, shallow sediment laid down during the recent Holocene settles like fresh soil in a garden. Deeper underground, buried Pleistocene-era deposits compact more slowly under greater pressure. Some 12 miles below the surface, where we finally hit solid ground, Gangetic basement rock sinks in isostatic adjustment to the ongoing rise of the Himalayas.

However, some parts of the delta have been settling faster lately. The water level in rivers has been rising for the same reason that land inside the polders has been falling. Seasonal deposits of fresh sediment used to accrete across the delta. This created new land and maintained its elevation above sea level. But now, that process has been stymied by the embankments—there is nothing to offset natural subsidence. Before the polders were created, the southwest delta was naturally subsiding by 5 to 7 millimeters a year, a rate easily matched by the piling of floodborne silt. However, since the 1960s, the polders have lost up to 5 feet of elevation, and recent models show that during storms like Cyclone Sidr, which claimed 3,363 lives here in 2007, this loss and the related lack of river drainage caused more land to flood.[29]

"Bangladesh is still growing in some places, but sedimentation is not evenly distributed," Steckler explains. "The poldered areas are running into problems."

This was obvious as Quiah and I rode along Pratapnagar's rough brick roads. Even at low tide during the dry season, the farmland on my right sat at least 10 feet below the waterline on my left. And all that stood between them was a crumbling mud wall barely wide enough for two motorcycles to pass.

During the last five decades of intensive engineering under the Coastal Embankment Project and other schemes, the delta's technologically improved lands have dropped faster than rising seas have consumed the coast. As Wilson writes, "It is becoming increasingly apparent that direct human manipulations in naturally dynamic deltas and their watersheds have to date had a substantially greater impact than climate change or global sea-level rise."

In effect, Steckler, Wilson, and the other monks have pieced together a very large elephant in the room, because embankments and the polders they contain are exactly the strategy Bangladesh is now using to save itself from climate change–driven sea-level rise. As the *Kokilmoni* follows the Pasur River downstream out of Mongla, contending with a 13-foot tidal flux that exposes formidable midriver sand bars, sailing through fleets of enormous ships laden with massive globes of natural gas, passing curious farmers who wave from the banks, and entering the dizzying maze of tributaries that coil through the Sundarbans, the question I keep asking is, Why?

Water and Power

Quiah and I are in the far southwest corner of Bangladesh. We've pulled over to a cha stand covered by palm fronds to take a break from the hot and ragged road. A narrow river slumps along like a thin milkshake below us. I'm sitting at one end of a wooden bench, enjoying a rare, momentary break from sensory input. Quiah smokes contemplatively at the other end. Black crows bark at each other from a pair of spindly trees, and I watch the cha *wallah* work. This barista needs only eight small, glass cups, which he cycles through with rote precision. A man approaches and orders *doodh cha*, or milk tea. While he draws a cigarette from his shirt pocket, the wallah goes to work swinging tin kettles and stirring thick condensed milk into the red tea. As soon as the man finishes his cup, the wallah dips the used glass into a plastic tub of lukewarm water and lines it up for the next customer.

Quiah takes a drag, blows smoke through the stand's bamboo latticed wall, and asks whether I've heard the story of the Royal Bengal tiger. I haven't. There once was cat from Bengal who traveled to America to be a prize fighter, he says. The cat was fierce and fought well. Other cats came from all over the world to challenge him, and he beat them one by one. As his victories mounted, people took notice, and he soon became a celebrity. When he finally won the world cat-fighting championship, a reporter asked how a cat from a place like Bengal could have triumphed over so many stronger felines. He replied that in truth he was a Royal Bengal tiger, but famine had made him a cat.

The polders and embankments that we've been riding on and that collapsed around Pratapnagar appeared here in the latter half of the twentieth century. They might be mistaken for modern works, but in truth they date back centuries to a people who began exploiting the riches of this place the minute they saw them. Quiah and I are on our way to a tiny village located just across a river from India to meet a man with roots closer to the beginning of this story.

The Portuguese sent the first European armada up the Hooghly River to modern-day Calcutta in 1530.[30] They were searching for a sea route to the region's coveted spices, trading partners, and locals to convert or enslave. They couldn't hold Bengal, however, and were soon elbowed out by the Dutch, who built their own trading forts from the Hooghly River to Dhaka. Then came the English. Then the French, one of whom, a Parisian merchant named Jean-Baptiste Tavernier, wrote,

> Egypt has been represented in every age as the finest and most fruitful country . . . but the knowledge I have acquired in Bengale, during two visits paid to kingdom, inclines me to believe the pre-eminence ascribed to Egypt is rather due to Bengale.[31]

European traders ventured into these tiger-prowled forests and mosquito-infested bogs seeking pepper, cinnamon, coriander, and mustard.[32] It was the dawn of the global corporation, when merchants pooled their resources, outfitted ships, and raised armies to do business. So came da Vereenigde Oostindische Compagni, la Compagnie des Indes Orientales, and the English East India Company (EIC). They overwhelmed a disunited Islamic Mughal sultanate, and in the inevitable war that followed, the British prevailed. By the late 1600s, the EIC was tightening its grip on what it called Indostan, chipping away at the crumbling Muslim influence and fending off the pesky French, just as their countrymen were doing in North America.

"We must lay down one fundamental maxim with regard to Indostan," wrote an EIC officer in 1687, "which is that as we acquired our influence and our possessions by force, it is by force that we must maintain and preserve them."

After several boat crossings and more jostling on the back of a motorcycle, I arrive in Kalinche. Quiah and I are within 50 yards of the Sundarbans, which contain the last intact remains of a once vast deltaic forest. We park outside a small police office and have only been inside for a minute or two before I make the mistake of asking Gopal Munda whether his people are Hindi or Muslim. It's a stupid question—they are neither—but he graciously turns my gaffe into a teaching moment.

The Munda developed their own religion and language in the mountains of Jharkland, in northeastern India, where they lived until the British asked them to move south and cut trees. Gopal is their descendant.

A police officer with a long, clean-shaven face and starched royal blue uniform, he's been tracing this history. We talk in a whitewashed meeting hall with high ceilings, long wooden tables, and creaky chairs. Across the room, two policemen discuss a pile of paperwork, and a thin old man in a neat white *thobe* leafs through a massive book of birth records.

The Munda were not having an easy time in Jharkland, Gopal says. Then came the British. "They were looking for places where the people worked hard, and Jharkland had hard workers. The British contracted community leaders to find strong people who could cut down the jungle." The EIC viewed the delta's vast forests as inexhaustible wastelands and set about clearcutting them to make room for more profitable crops. Starting on the south side of the Ganges River, they chopped and sawed their way toward the coast. Timber was exported and land was cultivated. Once the forest had been stripped bare, the British offered parcels to those who chose to stay and farm. "This opportunity kept us here," Gopal says.

The EIC inherited and streamlined a Feudal Age system of governance from the Islamic Mughals in which locally based landowners, called *zamindars*, lorded over fiefdoms. The peasants' labor earned income for the zamindars, who paid a portion to the EIC as rent. Under this British corporate peonage and marooned in a clearcut jungle that was often under water, the Munda struggled to survive. Most only had one set of clothes, and they didn't know how to work the land, Gopal says. They survived on tree roots and lotus flowers and kept pigs for a while, but they gave them up when Muslim workers joined their villages. Despite English promises, the suffering of Jharkland appeared to have followed the Munda.

This did not bode well for the rent-paying zamindars. To maximize profits, they pressured the farmers to replace rice that could be eaten locally with cash crops like cotton and jute for export. Tea arrived in Bengal in 1835, after Europeans grew tired of paying for the Chinese product.[33] For the Munda, this transition was disastrous: If they weren't growing food, who was? And the zamindars' problems only worsened as the peasants starved.

Saddled with the crippling weight of this system, the zamindars built some of the earliest embankments. People native to Bengal had long protected their crops by piling low, earthen *bhunds* between them and the river. These temporary berms were purposefully breached each monsoon season so the land could drain. However the English and the zamindars jacketed stretches of rivers with permanent barriers.[34] It was a new solution to a new problem: The embankments were necessary only because the delta had been clearcut and converted to farmland.[35]

It worked at first. Raising riverbanks kept water in its channels, protected farmland, and lengthened the growing season. Soon, though, ominous signs emerged. On land, crops withered and starved for nutrients that seasonal floods used to bring. In the rivers, channels filled with sediment. They rose and cut into their banks, tearing down the feeble mud levees and unleashing floods like the people had never experienced. Desperate, the peasants fell back on foraging in the mangrove forest they had helped to clear.

"Neither the *Zamindars* nor British Engineers could foresee the hydro-geomorphological impacts of embanking the rivers in the long run. The sense of protection against flood ensured by the embankment ultimately proved to be myth," writes Kalyan Rudra, chairman of the West Bengal Pollution Control Board in Kolkata. "The temporary benefits derived at the first phase gradually disappear, and the cost of ecological degeneration was externalized on the next generations."

Those costs are still being paid today. What remains of the Sundarbans is protected as a United Nations Educational, Scientific and Cultural Organization (UNESCO) heritage site that stretches across the border with India. Gopal helps to manage it. He volunteers to protect locals from tigers that wander into villages. He says this nonchalantly, perched on a stool with a hand on his hip. Managing the Sundarbans also means chasing thieves and poachers who paddle into the tiger's territory to fish, hunt spotted deer, and cut trees. With so little of the forest intact, and even that under pressure from pollution and climate change, the Bangladeshi government has restricted access to allow vital fish and crabs a fighting chance at breeding. Enforcing these rules has put Gopal at odds with his own community.

It's a huge loss for people scraping by, he says. And lately, the Munda have been on the move again, leaving Kalinche for work in northern brick factories, not unlike their ancestors who went cutting trees.

Satanic Chains

The first time I visited Bangladesh, I landed at the height of a historic outbreak of dengue fever. Rural hospitals, which are enough to dissuade sickness under normal conditions, overflowed with feverish victims. Patients spilled out into open-air breezeways where they sprawled over woven mats on concrete floors. In the dengue ward of a hospital in Khulna, the sickest writhed under pink mosquito net tipis while family members sat by quietly to comfort them. A doctor told me he had never seen it so bad.

In dense city neighborhoods, government workers sprayed thick white clouds of insecticide while gangs of giggling children danced in the roiling fog. I began each morning by slathering myself with DEET and reapplied unhealthy coatings of the cancerous stuff all day. I'd already been vaccinated against hepatitis, tetanus, encephalitis, typhoid, yellow fever, and malaria, but there's no vaccine for dengue, which is spread by mosquitoes. A year before the outbreak of COVID-19 sent the world into paranoid lockdown, here was a taste of the haunting anxiety that makes one doubt the air around them.

The delta's pestilence and fear of it haunted the souls who toiled for the EIC. For more than half a century, the company benefited from a nearly unchallenged monopoly on business in the Indian subcontinent, and it used the privilege mercilessly. Under EIC rule, disease was rampant, infrastructure was lacking, famines were regular, floods were devastating, and, always, the peasants suffered.[36] To German economist Karl Marx, the ruinous machinations of British export capitalism had turned the idyllic Indian subcontinent into the Ireland of the East, "a world of voluptuousness and of a world of woes."[37]

"England has broken down the entire framework of Indian society, without any symptoms of reconstitution yet appearing," Marx wrote in the *New-York Herald Tribune* in 1853. "This loss of his old world, with no gain of a new one, imparts a particular kind of melancholy to the present misery of the Hindoo, and separates Hindustan, ruled by Britain, from all its ancient traditions, and from the whole of its past history."

At last, when the British Crown was forced to revoke the EIC's monopoly and Bengal opened to free trade, European enterprise molded the region from the Indus River to the Chittagong Tracts into a vast work colony tuned to supplying raw materials for English industries while stifling any opportunity for Bengalis to prosper. The Industrial Revolution drove this exploitation to a fever pitch on the back of its throbbing incarnation: the steam-powered locomotive. While London's factories churned materials into products at breakneck speed, trains carried goods and soldiers farther and faster across Bengal.

Where trains went, trees fell and stagnant wetland prevailed. More forests were cleared to make railroad ties, or sleepers, and rail lines were laid atop embankments made of mud excavated from ditches beside the tracks. These shadow ditches became standing pools while the elevated lines beheaded rivers and streams, altering natural flows and creating wide, sluggish bogs. Mosquitoes loved British infrastructure. Malaria was rife. A mile of track required about 1,700 sleepers, and the malarial costs of some stretches were said to be "a death a sleeper." Bodies were left to rot along the way. This abhorrent misery prompted Willcocks, who coined the term *dying rivers*, to dub the railways "veritable Satanic chains."[38]

By the end of the nineteenth century, the British Crown had spent 150 million pounds sterling on railways across India, its single largest investment of the time.[39] More rail brought more embankments, which meant more standing water, more disease, less sediment spread and so less natural fertilizer, fewer crops, and more hungry peasants. Native Bengalis resisted quietly, cutting embankments to drain the land. Somehow, though, it never occurred to British engineers that these breaches had been on purpose.

"Have you had any experience with the effects of drainage on waterlogged land?," the inspector-general of agriculture for India was asked in 1901.[40]

"No, I should like to try that experiment," he replied.

He didn't.

Meanwhile, Bengalis such as chemist Prafulla Chandra Rây called the British government "criminally and willfully responsible" for this havoc. "The fact is that railway lines are always constructed with an eye to the interest of foreign shareholders. The less the cost, the greater the expectation of dividend," he wrote. Rây was born in India in 1861 and educated in Scotland. He returned home to found a pharmaceutical company, write a *History of Hindu Chemistry*, and organize relief for famine and flood.[41] He saw that the British attempts to control or eliminate flooding in the Delta contradicted the native Bengalis' flood-adapted habits. They had learned to live with floods. They even named their daughters after them. Like snow to the Inuit, floods were not all the same, and only some were disastrous.

Bengalis have names for the good floods, *barsa*, that come during the rainy season and replenish dry fields; bad floods, *bonna*, brought by storms that cause destruction; and frustrating floods, *jalabaddho*, caused by waterlogging from poor drainage. The British didn't understand this. They were busy laying track.

~

By the time the Crown finally retreated from the Indian subcontinent, in 1947, eastern Bengal was in shambles. And as a parting shot, the British Raj threw oil into the fires of a long-simmering religious conflict. When India gained independence, the subcontinent was partitioned into Hindu India and Muslim Pakistan, which consisted roughly of the Pakistan we know today and a satellite East Pakistan that eventually became Bangladesh. Atop this seismic shift in foundation, old sociopolitical tensions erupted in horrific violence: Muslims marauded in India, and Hindus cut off in Pakistan took flight in one of the largest migrations in human history. They cut each other down by the millions as they ran for the border, so that train stations were smeared with blood.

Independence in India did not mean independence in East Pakistan, however. Bengalis remained under the yoke of an absentee ruler based in western Pakistan. They were still overrun with haphazard infrastructure, still under water, and still hungry. Such was the context that Julius Albert Krug dropped into in the 1950s.

Krug was an American engineer whose career had been defined by water and power. He was baptized at the Tennessee Valley Authority. Later, as secretary of the interior, he signed off on the construction of the Grand Coulee Dam in Washington State and oversaw the development of a string of dams along the Columbia River, as well as some of the largest undertakings on the Colorado River. In 1955, the United Nations (UN) called him to East Pakistan to find a solution for the 45 million people who were mired in devastating floods.

From an airplane in the dry season, the delta's rivers look like strands of silver hair fallen over a green carpet. When Krug visited, the swollen waters would have looked like wide rice noodles encircling tiny islands of green. A third of the country was inundated. Thousands of people had been killed, and tens of thousands of square miles of mostly farmland had been ruined for the second year running.[42] Fleeing the countryside, mothers snatched their children, roped their cattle, and set off for safety in nearby cities. They crowded onto roads and huddled in parks en masse with nothing but what they carried. Men stayed behind in the pouring rain to build futile mud dams around their homes. It was called the flood of the century. *The New York Times* bemoaned it as "a financial catastrophe for the peasants."

A disciple of large, centralized infrastructure, Krug took a look around the delta and recommended that Pakistan throw its whole weight behind a massive effort to "cordon off" swaths with high embankments modeled after the Dutch polder. The East Pakistan Water and Power Development Authority was created to manage this effort, and it began drawing a blueprint for a hydrological revolution that would change Bengal forever.

America of the 1960s was caught in the throes of the Green Revolution and a jingoistic belief in salvation through technology. Post-Partition Bengal had been scrolling by in Western newspapers as images of children looking like skeletal pears with nothing in their eyes but the singular want for food. The West had solutions: tools and expertise honed during its own trials with unruly waters from Tennessee to Arizona. It would bring these to Southeast Asia to face two familiar foes: flood and drought.

In northern Bengal, where tea was grown, wet and warm tropical air mixes with hot air rising off the land, stalls when it collides with the Himalayan highlands, and creates the monsoon. These short but torrential downpours produce punishing summer floods. Krug reasoned that they needed to be captured and diverted for irrigation in dry times before bleeding away. The coast is different. Here, the rivers are tidal.[43] Twice a day, the ocean breathes with a force felt 200 miles upstream, while water at the mouth of the bay rises and falls by 10 feet. When the ocean turns angry with late season cyclones, storm surges swell rivers and topple their banks with battering-ram waves. This water needed to be resisted.

The Pakistani water and power authority took Krug's advice to heart and designed more than one hundred Dutch-inspired polders to protect 3 million acres.[44] Work had just begun when another massive flood prompted the UN to send another envoy, this time headed by John Ray Hardin, a former president of the Mississippi River Commission, to check Krug's math. "After a century or more of progress, the art of river engineering has, to a large extent, become a science," Hardin proclaimed, apparently satisfied. The year was 1963.

Locals remained unconvinced. Prasanta Mahalanobis, a professor from Kolkata, had warned as early as 1927 that a flood control strategy focused on large embankments would ultimately fail in Bengal.[45] This Western scheme overlooked the fact that coastal Bengal would not exist without the sediment deposition of floods. Mahalanobis warned that containing the rivers would cause them to clog and the land to sink. But the UN dismissed this warning, retorting that the effect of embankments on sedimentation would be small and localized. The Coastal Embankment Project began building polders.

All the while, a new tension stressed the relationship between the divorced Pakistans. Despite having less than half the country's total population, West Pakistan hoarded twice as much money for its own flood infrastructure as it gave to East Pakistan.

"What kind of justice toward East Pakistan is this?" cried Bangabandhu.[46] Islam and a shared distrust of India had kept the sides tethered, but their bond was fraying fast. When West Pakistani leaders decreed that Urdu, a tongue spoken in the west, would be the national language, a Bangla-speaking revolution erupted in the east.

Eighteen days after Bangabandhu's famous speech, Pakistan Armed Forces soldiers slipped into Dhaka University under cover of darkness and attempted to stifle the liberation movement by executing professors and students who were championing Bangla culture. It began a months-long genocide. But rather than stamping out Bangla, the Pakistani onslaught rallied tens of millions of people in revolutionary fervor for self-determination. In December 1971, after more than 3 million deaths, that movement delivered an independent Bangladesh.

War did not solve the problems of drought and flood, however, and in some ways it made them worse. It cost the new country innumerable resources and some of its nimblest political and scientific minds. Limping along in its aftermath, Bangladesh sought new benefactors, and they materialized under modern banners: USAID, the World Bank, l'Agence Française de Développement. In the years after the revolution, these and other international organizations stacked their donations behind crop production and flood control in the delta, though, as would soon become clear, not with purely altruistic motives.[47]

~

By the 1980s, the seeds of Krug's vision had been sowed and the influence of a new foreign authority, in the form of international aid, was well entrenched. August 1988 brought one of those floods that people talk about for generations. Monsoons topped rivers in the north while a cyclone pressed the roaring ocean against them.[48] Sixty-six percent of the country was inundated. Even Dhaka, which was thought to be protected by the massive Dhaka–Narayanganj–Demra embankment built as part of the 1960s-era Coastal Embankment Project, sat under water for weeks.[49]

The wife of French president François Mitterrand, who was in the city at the time, got her feet wet watching people pile sandbags night and day for almost a week. Back in France, Danielle leaned on her husband to rescue Bangladesh. He sent engineers with a solution—the Flood Action Plan, which sought to make Bangladesh "impregnable to flood" by augmenting the Coastal Embankment Project through the construction of nearly 2,500 miles of massive embankments. This solution was expected to take twenty years to complete at a cost of $10 billion.

The United States offered its own solution, but it took a curiously conservative tack. Its lead author, a Harvard engineer named Peter Rogers, emphasized the need to "live with flood" and even recommended against building more embankments. This approach reportedly infuriated the Bangladeshi government, which had its sights set on the multibillion-dollar option and found the American proposal "defeatist." Large-scale infrastructure was not only considered the best defense against the delta's punishing hazards; it also promised to bring billions of foreign dollars into the Bangladeshi economy. Local consultants, construction firms, politicians, and, to a much lesser extent, local laborers all had something to gain from these monumental public works projects. And yet their profits paled in comparison to what outside investors stood to make because, like schools of salmon, foreign aid dollars reliably swim home to their source.

"Consultants like USAID have commitments to their constituencies at home," said Kimberley Anh Thomas, a geographer at Temple University and an expert on the politics and economics of adaptation in the developing world. Eventually, aid projects must answer to the American taxpayer and prove that their work benefits the American public. "The way that you demonstrate that is by mandating that materials, supplies, and equipment have to be sourced from the US. In US dollars."

In late-twentieth-century Bangladesh, a country reeling from decades of famine and genocide, USAID charged interest on loans it provided to construct embankments.[50] It also mandated that US consultants be employed for the work. Even a corn farmer in Iowa had something to gain from adaptation in Bangladesh, because of a US program called Food for Peace, which offloaded surplus American crops to developing countries as a means of insulating US growers from depressed markets. The risks "were displaced onto the rural peasantry of Bangladesh, who suffered debilitating food price fluctuations as a result of imported foodstuffs," Thomas found. This relationship continues today. Charities, companies, and governments in developing countries see less than 10 percent of the US government's $35 billion annual budget for foreign investment. The majority of funds are funneled back to Washington-based companies and nongovernment organizations (NGOs), commonly known as the "Beltway Bandits."

Often, investment ends with construction, so that locals must maintain complex infrastructure on their own and stomach its unintended consequences. During the devastating 1988 flood, for example, damage was worse inside the embankment built along the Meghna River, partly because water had been trapped inside the polder and partly because the presence of the walls had encouraged residents to abandon the practice of building on elevated platforms. As I had seen in Japan, a cultural instinct rooted in generations of self-preservation and local knowledge had been overshadowed by a new trust in alien technology.

In 1990—the year construction under the Flood Action Plan began—an internal audit at the World Bank acknowledged an "extraordinary absence" of evaluations of existing flood controls where new walls were being planned. Containing the rivers hadn't worked so far, the scathing review found, and yet there was still "continuing pressure for large scale capital intensive 'solutions' to the flood control problem when all available evidence indicates that such schemes have not been cost effective in the past and are unlikely to be in the future." In the end, however, Bangladesh's Flood Action Plan, which was coordinated by the World Bank and paid for by donors including the French and Dutch, forged onward.

"The British left India in 1947 but the old philosophy of command and control over the nature continues," Kalyan Rudra writes.

We Do It Ourselves

Quiah and I say goodbye to Gopal, placing our right hands over our hearts in the local custom, and go searching for another boat to take us north. Ferry timetables run on some mysterious mathematical combination of the tides and a boatman's whimsy, so we wait. Joining us at the dock is a man on a bicycle-drawn cart with a cargo that buzzes. I peer over to see his burlap sacks trembling with hundreds of brown baby crabs that he harvested from the river and intends to sell to aquaculture farmers. Quiah leans against our parked motorcycle, a cigarette in one hand, his cell phone in the other.

So, it turns out, Bengal's dying rivers were not some brilliant scheme from the Levant. They are instead proof that we were born to a fickle mother. Clearly, though, the people who have lived with them longest have found ways to endure. To see how, we head inland for a place called Beel Dakatia. The poldered island lies a few klicks west of Prof. Datta's Khulna University office and north of Pratapnagar. It was also ground zero for a revolution in delta water management that began when its people risked their lives to cut a hole in their own embankment.

A *beel* is a saucer-shaped depression where water naturally pools, and at 75 square miles, Dakatia is the country's second-largest such place. After it rose out of the delta from accumulating silt, people built their homes around its raised edge. I follow a winding dirt road that zigs across bucolic paddy farmland and zags through quaint rural villages. Living some 80 miles from the coast, people here are less worried about cyclonic waves than they are rains that turn the beel into a quagmire. For months each year, dry fuel is a rare commodity, so women prepare dung sticks that they set out like lean-tos to dry in the sun. They mix cow dung with straw into hand pies that they slap to short wooden poles like predigested satay on skewers. Each leaves a unique handprint in the drying manure.

It's shaping up to be a hot afternoon in a village near Dakatia's center. Liberated school kids chase each other across a dusty yard with ecstatic whoops. Farmers saunter in from their fields heading toward *dighis*, small ponds carved into the mud behind homes. These ponds capture flood or monsoon water for drinking, cleaning, fishing, and bathing. Today, farmers arrive with dirt-caked sandals and bars of neem soap. Clothed in lungis, they crouch at the dighi's edge and lather their heads, arms, and chests with a thick white foam. Women join them, stepping into the dark pools and ducking under, so that their vibrant saris float up to the surface around them like blooming lotus flowers.

To live with the rainy season, Dakatia's earliest residents crisscrossed their farmland with narrow berms just a couple feet high. These demarcated one plot from another while providing a semidry path to walk. During high floods, when even these *ails* were sunken, people got around by boat. They planted fruit trees and delta-adapted crops, including a variety of grains, melons, and vegetables in patterns that took advantage of both the long stretches of aridity and the bursts of sogginess. In early spring, they seeded the miraculous *bona aman*, a native rice varietal that kept pace with rising flood waters by growing at a rate of 1 foot an hour toward an ultimate height equal to the average giraffe.[51] In late spring, they spread *aus* rice seeds that they harvested near the end of summer. During the dry *rabi* winter season, they sowed *boro* rice, potatoes, and lentils that thrived on only the irrigation farmers carried to them in tin gourds.

They built embankments, too, but not to prevent flooding. Their *bhunds* were temporary earthen barriers built at the beginning of the dry season to keep salty river water from poisoning their crops.[52] These "eight-month embankments" washed away sometime around June with the rainy season's onset, leaving the land open to the deposition of fresh sediment.

"They were intuitively following the principle of 'least resistance,'" writes Nazrul Islam, the United Nations chief of developmental research.[53] "They knew the roads and dykes obstructed free passage of water and aggravated floods," so they built few of them. "They knew that rivers gave birth to this land, and rivers would come periodically to nurture it. They realized that it was in their own interest to let this nurturing take place. Therefore, they struck a bargain with the rivers: instead of trying to prevent river inundation, they made best use of it."

I don't mean to give the impression that this was some Shangri-La. Even when welcomed, floods took their pound of flesh. So when Westerners showed up with plans for strong embankments that promised to turn Dakatia into an easily managed breadbasket, locals were happy for the help. In the 1960s, Beel Dakatia became a polder under the Coastal Embankment Project, and for almost a decade the scheme appeared to work.

Midafternoon is a good time to wander into a rural Bangladeshi town looking for someone to talk to. I find Kumaresh Haldar, or rather he finds me, and he is eager to help. I'm often struck by how readily locals drop what they're doing to speak with me, but then I think of how quickly I would nix my afternoon drudgery if Haldar appeared in my town, walking down Main Street in a plaid red lungi and bare chested except for a gold locket tied to a string around his neck. He leads me up wooden stairs to a yet another cha stand perched over a large pond across from the school and cyclone shelter. I take off my sandals, awkwardly as always, and duck into a large open-air veranda.

Haldar hears my question about the years leading to 1990 and nods knowingly. "*Ha, ha*," he says in the affirmative; "*pani bondir somoy*." The waterlogging time.

Back then, 165,000 people lived in thirty-six villages across Beel Dakatia.[54] When salinity from tidal rivers curbed rice and fruit tree yields, the government encircled the land in embankments and installed a sluice gate for drainage. The crops did well behind the walls for a time, but in the rivers, sediment was quietly piling up.[55] Then, in 1982, the monsoon rains pounded them and lingered.

"The land was down and then it filled. Everywhere there was waterlogging," Haldar recounts. The polder's few sluice gates, installed for drainage, had clogged with silt, and their heavy metal doors would not budge. "We complained repeatedly, but the government took no action," Haldar tells me. Weeks of waterlogging turned to months, and then years.

Dakatia's orchards became graveyards of pickled mango, raintree, betel nut, and coconut, and its fertile paddy was rendered sterile. "During the waterlogging time, our earnings from daily labor were less than the cost of rice," Haldar laments. "Many people left."

The community was engulfed in suffering, drowning in troubles that seeped into every aspect of their lives: food, water, income, housing, transportation, school, playgrounds—even their religious rites. With no dry lumber for pyres, the dead were left in the open, denied even the dignity of a proper Hindu cremation.

When a Bangladeshi economist named Atiur Rahman visited Dakatia in the ensuing years, he was struck by the psychological toll waterlogging had taken on residents he interviewed. Most of them remained only because they were too poor to leave.

"Santosh Biswas is not a quack," Rahman wrote of one Dakatian. "He has a degree in homeopathy. But he has been forced to give up his medical profession as it did not yield enough income to support his family. Now he is a part-time fisherman. He and his wife catch fish in Beel Dakatia in the early morning so that people don't see them." Asked what Biswas blamed for his desperate situation, the doctor-turned-fisherman replied, "When people used to live in harmony with nature, crops were produced and people could live well. The moment we constructed embankments to lead a better life we started annoying the nature. Our sufferings originate from interfering with nature. Embankment is responsible for all our sorrows."

~

It's painfully ironic that so many people starved behind the walls, given that the embankment building spree of the late twentieth century was about shielding investments in agriculture. The French, Dutch, Japanese, and Americans had tried to manifest, despite the physical reality, the environmental conditions needed for the wundercrops of their Green Revolution to outdo breeds that had evolved in the delta. Within the polders, diverse and flood-attuned native cropping patterns were replaced with intensive cultivation of high-yielding varieties of boro rice, which were entirely dependent on irrigation.

There's no denying the benefits that the Green Revolution brought to Bengal. With help from increased cropping, fertilizers, and pesticides, population soared from around 50 million in 1950 to close to 103 million by 1990. But development of this kind, on this scale, also produces subtle, long-term impacts. Under the surface, it molded the landscape and the local psyche with a tacit logic.[56]

Once set in motion, the Western development machine cannot help but find ways to apply itself. Wherever it goes, its consultants and experts produce plans based on the best available science yet typically skirting any local knowledge that threatens to challenge its programmed procedures. It functions best in clean settings, narrow confines, with no alarms and no surprises. I witnessed something similar in Japan, where the complexity of the Tohoku coast was simplified so that a singular solution—seawalls—could be applied universally. In Bengal, native arrangements with flooding were too knotted up in the delta's confounding seasons and shifting moods. The Dutch polder offered a clean slate suitable to the tool of irrigation. And by bringing the delta under control with this tool, the Western development machine consolidated, centralized, and co-opted the power to wield it.

In Beel Dakatia, locals grew dependent on Western varieties of high-yielding rice and infrastructure that was never theirs. By the time of the great flood of 1988, people lacked the money, time, and expertise to fix their problems. As he explains this, Haldar shifts to his hip and leans back on a straightened arm. Before independence, when the land was owned not by the state but by zamindars who had an incentive to keep the land productive, the complaints might have been heard, he says. In 1990, though, "We informed the government, but nothing happened. So we said, 'We'll have to do it ourselves.'"

At nine in the morning of August 13, Dakatians of all ages rallied at a school playground. Organized into local samities and whipped up by communist party leaders, they set out to cut parts of the southern embankment, reestablish normal tidal flow, and let sediment rebuild their beel. Most carried woven baskets and spades. Some played traditional drums and pipes. "Thousands and thousands of people came from all over," Haldar recalls. As they gathered into a throbbing mob with a singular purpose, law enforcement waited in the wings. "The police were told to shoot on us, but when they saw the huge amount of people they didn't have the energy to do it."

Through a steady rain, the mob gouged at the wall of gray mud. Their soaked clothing clung to their skin, and their feet slipped out from under them. They cut well into the afternoon, until they finally breached the embankment in four places. Then they stood back under the clearing sky to watch their efforts at work. Seeing Haldar glow as he describes the bravery of the people, I can only imagine how painful it must have been when the water rushed in, not out.

At the time of the cuts, the river was too high, its bed too silted up for the sunken beel to drain. Instead, the breaches inadvertently made things worse inside the polder. The moment might have been remembered as a tragic failure if the people at Dakatia had only wanted to drain the land, but they were also making a statement. All across the region, impoverished farmers were tired of being told how to live and then having to suffer the consequences. Cutting the embankment was a political act.

In explaining this, Haldar beams and recounts how the local chairman, communist party members, and farmers' rights activists traveled from Dhaka to march with him. On that rainy day, peaceful revolt, and the patient logic of deep knowledge had overwhelmed the will of centralized government. Bureaucrats took notice. Over the following years, the Water Board shored up embankments across 350 square miles and got around to dredging the river outside the polder to remove the clogging sediment. When that finally happened, it was as if the cork had been loosed from a bottle. Dakatia drained at last.

Meanwhile, word of the uprising leapt up riverbanks to neighboring polders, where farmers had also begged the government to do something about waterlogging and had also been waiting. With what money it had, the Water Board went on dredging, but as long as the embankments were in place, the rivers only silted up again, sometimes just one season later.

In the early 1990s, the Water Board hired Dutch and Australian consultants with money from the Asian Development Bank to come up with a long-term solution. In the Hari River basin near Dakatia, a slew of options simmered down to just two: a permanent tidal reservoir near the top of the channel, where the river could overflow and drop its silt, or a multi-million-dollar concrete barrage that would span the river downstream to form a defensive line against sediment-laden tidewater. Unsurprisingly, the board set its sights on laying a huge piece of concrete across the delta.

Then, in 1997, villagers in nearby Beel Bhaina who were fed up with technosolutions and feared that a barrage would cause clogging farther south took a page from Dakatia's playbook. They cut the embankment holding back the Hari.[57] Again, the communist party had egged them on and again the government sent police to quell the uprising, but this time a small number of foreign engineers and consultants intervened. What would happen if the cuts were left in place, they asked? The Water Board agreed to wait and see.

The experiment paid dividends. Over the following four years, the Hari hauled enough sediment into Bhaina to fill 2,600 Olympic-sized swimming pools and raise the land in some places by six and half feet. Just as important as the flood tide, though, was the ebb tide's impact. Like water through a tightened nozzle, the retreating river was yanked out of the narrow cuts faster than it had been pushed in, and it scoured the long-choked riverbed. The high-velocity outflow gradually carved the shallow river into a deep and swift-flowing channel. It was free dredging.

Before long, villagers in a nearby polder cut another embankment. And then another. The flint stroke at Dakatia was throwing sparks, and everywhere they caught, Bangladeshis were working out the kinks in a novel technique of living with tidal rivers.

The Lungs of the River

"I want to make you understand," declares Jahin Shams Sakkhar. He's beaming with a wide, boyish grin. "I am saying that we cannot be poldered anymore, right? But you cannot completely de-polder, because that would mean a catastrophe—that would mean a humanitarian crisis. So, what we can do is *controlled* de-poldering."

We sit on either side of a wooden table in a small office 15 miles southwest of Dakatia. The sun has just set outside. An overhead fan threatens to loose itself from the ceiling and, despite two mosquito coils smoldering away below the desk, the damned vectors strafe us in unceasing waves. Sakkhar has a frenzied look in his eyes. His black beard is thick and unkempt; his curly, dark hair is disheveled. He's sweating through a mustard-colored t-shirt, not from the labor of his breathless explanation of fluid dynamics but because until a few minutes ago he was outside playing cricket. The game may be his favorite thing, but talking about the delta is a close second.

"I don't mean open the whole polder, because some polders are huge. Like, for example, the polder currently we are living in is actually comprised of almost a million people. But we can open one tiny portion, around 2,000 acres of land," he explains.

Sakkhar is the son of Shahidul Islam, a respected delta researcher and the director of a locally operated NGO called Uttaran. He inherited his father's drive to help his impoverished country and added a master's degree in climate change and international development from the University of East Anglia in England. Uttaran set up shop as a human rights organization in 1989. It stood up for the peasant class of farmers who lived and worked on government-owned land because they were too poor to own their own. Inevitably, that work involved lobbying the government to invest in natural or nature-based solutions, waging a relentless resistance to techno-infrastructural answers that perpetuated a colonial mindset.

Sakkhar isn't old enough to remember the cuts at Dakatia, but he has strong opinions of the measures that were proposed before the people took charge. "The government plan was actually to stop the river," he says. "So silt is coming, 'Let's cut off the silt!' It's like your head is hurting so you cut it off. That's, like, a very stupid idea." Instead, Sakkhar, through Uttaran, is a fierce proponent of the strategy that was born in Beel Dakatia, which has come to be known as tidal river management (TRM).

"TRM is basically the lungs of a river. Our lungs expand, they take all the poisonous things out of our system. The lungs are like a filter, right? TRM acts like a filter for the rest of the river. It eats sediment and allows the delta to form." The method is often called an indigenous practice, because it uses something those intimately familiar with tidal rivers have known for a long time. "The monsoon is not a concern," Sakkhar assures me. "The concern is during the dry season, because a single river can carry 1.4 million tons of sediment in February, March and April. So all you actually have to do is manage these 1.4 million tons of silt for only three months. It's as simple as that."

Well, nothing's ever *that* simple. The trick with TRM, the lesson it has to offer adaptation efforts all over the world, is that it's place specific. Whereas the Dutch took a strategy of managing water that worked well in Holland and force fed it to riverside people across Asia, TRM was born in the dying rivers of southwest Bangladesh. It won't solve the droughts plaguing the northern Barind Tract, nor will it solve sea-level rise on the eastern coast where the Meghna River enters the bay. But here, it shows great promise.

Sakkhar leans his wide frame across the table and tears a page from my notebook. He scribbles a diagram, filling the blank space with a jumble of numbers and arrows. In his illustration, TRM starts with two parallel lines that represent a river running north–south. Next he draws a circle beside the river; this is a small, temporary tidal basin that the river will spill into. A low embankment modeled off the earlier *bhunds* is raised in a ring around the basin to contain the incoming water. He then notches a short, narrow canal that links the river to the basin. Finally, a temporary dam is erected across the river just above the canal to force its water into the basin.

The tide rushes upriver, runs headlong into the cross dam, and follows the path of least resistance into the TRM basin. Here, it stagnates just long enough to drop its sediment. Hours later, the outgoing current sucks the river back through the connecting canal and dredges the channel as it retreats to the sea. At each turn of the tide, the river leaves behind a new layer of silt, just like it used to, but in a more convenient, controlled fashion. After years, the basin will rise with nutrient-rich

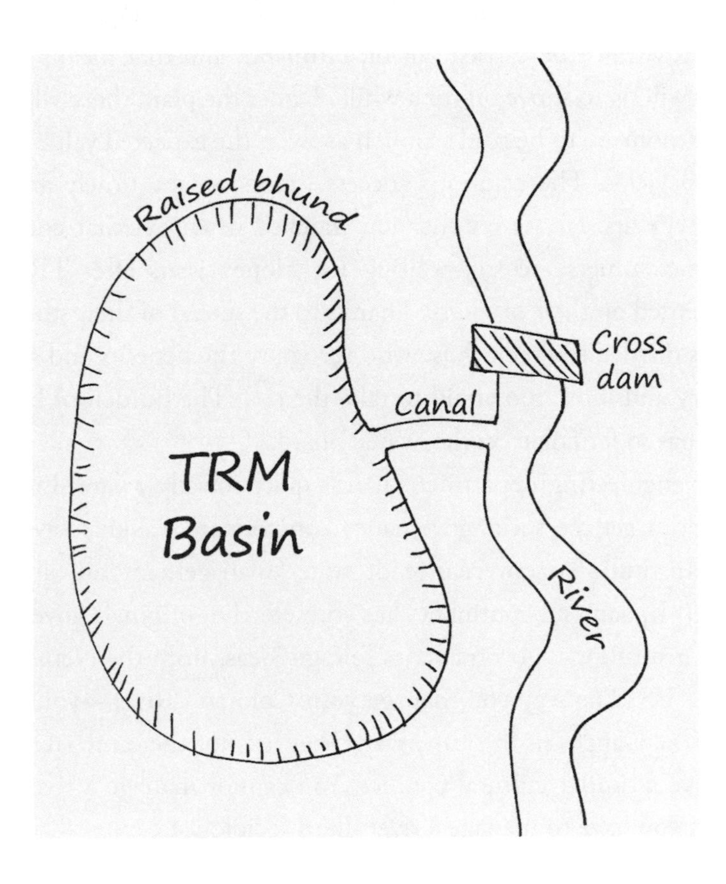

soil until it stands above the high-water mark. At that point, the canal is closed, farmers return to sowing in the raised basin, and the process repeats a few miles downriver.

Over time, the goal is to harness the fury of the dying rivers to build up the subsided delta and create productive farmland. "In that way, you do not just save your river, but you save your bottom from climate change as well," Sakkhar says with a smile.

He spins the paper around to face me and checks to see whether anything has clicked. "It seems so romantic, right? Everything is perfect. The engineering is perfect." He gives a thumbs up to the paper. "Yeah, it's absolutely perfect. But there's a catch: a governance challenge."

Since just about every acre of Bangladesh is claimed (legally or otherwise), TRM must be carried out on farmland, and that means farmers must be willing to shove off for a while. Under the plan, those who agree to make room are to be paid as much as twice the expected value of their sacrificed yields. The solution's success hinges on the timely arrival of those payments. I can't say that I'm shocked when Sakkhar complains that some farmers are still waiting for money years after TRM was implemented on their property. Thanks to the spread of their stories, the region is divided between those who recognize the benefits and are willing to try and those too afraid to take the risk. The burden of bridging the gap has so far fallen on the Water Board.

"Our engineering department is not quite socially aware. Engineers simply don't get the social governance component," Sakkhar says plaintively. "Institutional governance of water management on all of the southern Indian subcontinent has ruined the original governance, because institutional governance is foreign ideas, from the Netherlands, from the US. The way you manage your Colorado River—you cannot manage the Ganges in the same way. They are not the same river. They don't have a similar cultural context. You cannot manage a river like a scientist, you have to manage a river like a sociologist."

Sakkhar is one of those Bangladeshi millennials who straddles the line between village kid and urban elite. Guys from the village wear t-shirts emblazoned with phrases like "Just sleep" and "A man without beard is no man," and they wear them with zero irony. Village guys ride three to a motorcycle and smile with teeth stained red from chewing *qat*. They stare unselfconsciously with eyes that have grown accustomed to watching things happen around them. Sakkhar wears a Ralph Lauren tee with no haphazard signage. Though he's just in from cricket, he can't hide the part of him that's also just in from a foreign land. He has left the country—I can tell from his eyes, which have no time for staring. He exists in two worlds. It's a weighty position.

When Uttaran began pushing TRM back in the 1990s, prime minister Sheikh Hasina promoted the idea in her campaign for office, and the Water Board even helped develop it in several waterlogged polders. "There was good democracy back then," Sakkhar says. In the years since, the country has settled into its authoritative and resource-siphoning ways. In 2007, the military stormed into the office of Shahidul Islam, Uttaran's director and Sakkhar's father, and arrested him for publicly demanding the government adopt TRM. They beat him viciously, breaking both his legs, and held him in prison for seven months. When I visited, a steel door fitted with multiple large locks protected his office.

Today, "government projects are public stunts—huge money, zero outcome. Some of them are quite important, but because it's a public stunt, it's not being implemented correctly. And because of the lack of democratic space, it's very difficult to speak against it," Sakkhar tells me. Unfortunately, this makes TRM's price tag one of its tallest hurdles, because subsidizing some farmers and piling a *bhund* takes hardly any cash—not real cash, anyway—and the less it costs, the less alluring it is to politicians hungry for big-ticket investment.

Sakkhar has learned to pepper his TRM pitch with terms like "Indigenous knowledge" for the same reason he brings up climate change: It opens foreign wallets. "I don't think Bangladesh suffers from climate change," he says with a splash of bravado. "I think Bangladesh suffers from lack of management. And I don't think it's because Bangladesh is poor. I believe it's because of political reasons. It's absolutely because of political reasons."

A Bridge to the Future

On June 26, 2022, thousands of Bangladeshis gather to celebrate the opening of a bridge across the Padma River. It marks the culmination of days-long partying throughout the country. The size and magnificence of the Padma strike me like the Grand Canyon, and this moment is something like the opening of the Hoover Dam. Colorful banners and balloons decorate both banks. Prime Minister Sheikh Hasina Wazed stands beside pictures of her father, Bangabandhu, in a not-so-subtle show of dreams dreamed and realized. She unveils an inaugural plaque and announces the release of a commemorative postage stamp. "It is a historic moment for all the Bangladeshis to be proud," she says.

The Padma Bridge is a monument to the country's economic progress and a statement of entrance into the modern world. It's also the vanguard of heavy industrial development that's driving continued risky adaptation in one of the world's most vulnerable places.

The longest bridge in South Asia stretches 4 miles across the murky Padma. Before its completion, the land route from Dhaka to the southwest relied on a ferry crossing where cars backed up for miles. The bridge shortens the drive by at least three hours. It eases the flow of bricks, garments, and natural gas between supplier and seller. It provides Dhaka markets with freshly caught fish and boosts the supply of sacrificial cattle for Eid ul-Azha celebrations, after years in which more than 2 million such animals went unsold due to transport bottlenecks.[58]

The bridge took nine years to build and many more to plan, design, redesign, and fund. Engineers used the largest hydraulic hammer in the world, taking 30,000 blows to drive steel piles almost 400 feet into the soft riverbed. Three broke under the strain.[59] "The unique hydrographic features of the Padma River have posed the most daunting challenge," said the Chinese ambassador of the megaproject his country funded.

Bangladesh's own executive engineer was more modest, saying simply, "I am very happy to be part of Prime Minister Sheikh Hasina's dream project."

Three months before the opening ceremony, I'm chugging along the lesser Pasur River a few miles west of the Padma with a group of college-age environmental activists. This is the same channel I followed on the *Kokilmoni* out of Mongla. The activists are members of the Waterkeeper Alliance, which fights to preserve and protect rivers all over the world. And the Pasur could use some help. In 2012, its lower stretch was designated a sanctuary for Asia's only remaining freshwater dolphin species: the endangered Ganges River dolphin and Irrawaddy dolphin.[60] Like hilsa fish and mud crabs, dolphins have suffered from overfishing. Now, a rash of new industrial development along the river's banks threatens their survival.

Sitting next to me is Fatima Zannat, a twenty-seven-year-old aid worker with a cyclone preparedness program. She wears a lavender burkha, gold nose ring, and leather sandals. While we talk she reaches into a blue handbag and offers me cloves to chew on. (I check my breath.) When storms approach, Zannat speeds through villages with a microphone urging evacuation. She tells me awareness of the danger has improved, saying, "Before, hundreds used to die, but now it's only a few."

Our little boat grunts downstream past an enormous dredger working to clear sediment near the bank. It sucks gray delta mud from under the surface and shoots it in a high arc across the embankment onto land. This Sisyphean dredging continues nonstop so that fishing trawlers and cargo ships carrying industrial goods can reach factories and refineries.

The Padma Bridge is just one of at least seventeen megaprojects going up around Bangladesh. There is also a seaport, railroad, mass transit system in Dhaka, airport in Cox's Bazar, liquid natural gas terminal and pipeline, and nuclear power plant.[61] Some of the largest are happening right here along the Pasur. The most controversial by far is the Maitree super thermal power project, also known as the Rampal coal-fired power plant. It's being built 6 miles upstream of us and is what brought Zannat to today's protest.

"We are young, and before we die we want to do something for our community," she tells me. "If we don't do anything for the future, who will?"

It might seem odd for a country that's taking an unparalleled beating from climate change to be constructing its largest coal-fired power plant in 2022, but that's what Bangladesh is doing. While the Dutch plot a transition to net-zero emissions with river-bound rafts of floating solar panels, Bangladesh is trying to catch up to the twenty-first century. Rampal's still-white smokestack looms through the haze of the dry season like a lone minaret rising 952 feet above flat farmland. When finished, the 1,320-megawatt plant will transform 12,000 tons of sub-bituminous coal into electricity each day. To prevent ships laden with coal from bottoming out, the government must initially dredge 17 miles of river bottom, removing enough silt to bury Central Park 33 feet deep.[62] Bad news for dolphins. The spewing ash and coal dust are bad for everyone.

If Rampal were being built anywhere else, it might have gone up quietly amid the smog of Southeast Asian development. But it's being built here, on the bank of a major river less than 9 miles upstream of the Sundarbans. There's a reason Bangladeshis call the mangrove forest "our mother." It has provided for them from time immemorial, and it remains a stalwart natural shield against cyclonic wrath. The mangroves' unique ability to hold their roots while absorbing screaming winds and violent surges provides Bangladeshis with their preeminent natural coastal defense.

"People know the Sundarbans is saving their lives. They also know this development is harming the Sundarbans," says Noor Alam, a former Bangladeshi politician and the group's leader. The government has environmental protection programs but does little to enforce them. Among the most immediate threats is the risk of a tanker ship sinking in the Pasur, injecting its coal, oil, diesel, and whatever else into the riverine ecosystem. In fact, three coal-carrying vessels sank in this river in 2021 alone.[63] There's been scant research on the existing local ecology, and industrial development pushes through without environmental impact assessments.[64]

The plant was presented under the banner of creating jobs and modernizing the country, but Alam says "locals don't get these technical jobs, and local fishermen have no alternatives when the river is polluted."

Rampal is also not an isolated case. Sawmills, rice mills, shipyards, shrimp processing centers, cement factories, and liquefied natural gas terminals are popping up all around the Mongla area. The Padma Bridge has quickened their pace and opened doors to more money making in the sinking southwest. This could be a boon for local entrepreneurs—if the wealth isn't gobbled up by those with greater means.

Already, hundreds of investors have seized at least 10,000 acres of land that had supported local livelihoods.[65] Between 2012 and 2019, some 3,500 acres of mostly rice paddy and shrimp ponds were filled with sand to make foundations for industrial buildings. The owners behind these investments, many of them foreign based, wield outsized control over national budgets and policies. They have helped to make Bangladesh one of the most market-friendly countries in the world—a true capitalist haven.[66]

The $1.5-billion coal plant is a joint effort between Bangladesh and India, which critics accuse of investing in to ensure a buyer for its low-quality coal. Alam notes that India's more progressive environmental laws complicate constructing such a facility at home. We're only 38 miles from the border. So India will reap the economic benefits, and Bangladesh will be saddled with the impacts, he says.

Alam calls it the latest incarnation of resource colonialism in Bengal. "We don't need to wait for World Bank money anymore," he says. "There are many other loaners, and they don't care about the environment, human rights or democracy."

The boat slides up against a small dock on the sludgy west bank, and we disembark at a village pinned between the Pasur and the Sundarbans. We walk along an embankment that's been fortified with sandbags and join a group of about fifty women in a clearing. Most of them hold signs I can't understand. Clearly, they want to stop construction of Rampal, but when the protesters see my notebook they crowd around yelling complaints about outsiders who use poison to fish en masse and baby shrimp harvesting and the collapse of crab populations and the government's restrictions on gathering from the forest and the crumbling roads and the bad water and the fact that the nearest hospital is two hours away.

There's anger in their eyes, but I know it's not directed at me. I've stumbled into a tug of war. On the surface, it's between nature and industry, but really it's a contest between the economic development goals of the entire country and the needs of local people. What a mess we've gotten ourselves into when these aims take such divergent paths.

Then again, poor people are vulnerable people. If Rampal brings electricity that powers factories and creates wealth, it might buoy the coast against what's coming. Bangladesh makes only about $15 million annually exporting rice while spending $1 billion on climate adaptation. With a quarter of the delta at risk of being flooded by 2050, experts predict adaptation costs to quintuple. This presents an opportunity.

In 2018, Bangladesh adopted its first nationwide, comprehensive disaster and resource management plan since 1972. This joint production with the Netherlands, called Delta Plan 2100, aspires to do everything from alleviating poverty and providing drinking water to stabilizing eroding riverbanks.[67] The thread throughout, of course, is the impact of anthropogenic global warming, and Sakkhar is encouraged by early studies that acknowledged polders have played a role in southwestern woes, but the priorities of the $38-billion plan follow an all-too-familiar playbook: command and control.

Topping the list is the construction of two massive concrete barrages, or divergence dams, to regulate the flow of the Ganges and Brahmaputra rivers. There's some cognitive dissonance within a plan that purports to favor "no-regret" and "non-lock-in" solutions while suggesting construction of dams across the world's largest delta. Of the top nineteen priorities, seven call for building or updating coastal embankments, strengthened seawalls, or new polders. It should come as little surprise that a plan for adapting the delta as envisioned by the Netherlands would prioritize polders.

Local nonprofits like Sakkhar's Uttaran aren't the only ones peppering their mission statements with mentions of climate change. Huge contracts are at stake, and to win in the face of stiff global competition, donors must stand out from the crowd. To keep up with the evolving foreign aid landscape and to woo commitments from nations in need, governments and NGOs have rebranded their development agencies as development partners and reframed their work as locally tuned adaptation. Although the rhetoric has evolved, the solutions remain familiar and tuned to each foreign country's specialized area of expertise. The United States, which taught the world how to farm, promises to maximize agriculture sector growth through "climate-smart technologies." The Japanese are experts in disaster risk reduction. And the Dutch have convinced the world that they know how to manage water.

It's a classic case of hammers looking for nails, said Thomas, the Temple University geographer. "To make yourself relevant, you have to frame the problem in a way that makes you exactly what that problem needs."

As a result, the Bengal's myriad intertwining issues are often distilled into the direct results of climate change–driven sea-level rise, a straightforward problem that some still believe can be overcome with simple but expensive infrastructure.

In addition to the Delta Plan 2100, since 2013 the World Bank has invested in updating more than 80 miles of embankments as part of the Coastal Embankment Improvement Project (CEIP). Under this latest iteration of river control that dates back to the Krug mission, foreign donors are spending upwards of $400 million with the rationale that a bigger version of an old solution to an ancient problem constitutes adaptation to climate change. Hard infrastructure undoubtedly has a role to play in adapting the delta, but given what we know about their adverse effects, why do these projects continue to gobble up funding while softer and cheaper options like TRM stagnate?

TRM is challenged by the fact that the embankments are a result of past famine relief efforts that were "built by the Dutch and funded by the US," said an American geologist who has worked in the delta for twenty-five years and who asked to remain anonymous out of concern that he might lose access. Rejecting that strategy now would not only pose a logistical hurdle, he told me, it would mean admitting a gross misstep. It's easier to blame the delta's problems on climate change. Really, "It's more complicated," he told me, "but complicated doesn't bring in any money."

Framing every problem as a result of climate change flattens the world to a single dimension. It muddles history and severs lines of causation, narrowing our view of the challenge and shrinking the lens through which we search for solutions. Basically, it allows us to forget the past so that we might make the same mistakes again. This process of reductionism helps explain the struggle over brackish water shrimp farming that engulfed Karunamoyee Sardar and the others in Polder 22. As it was framed by foreign governments and NGOs, seawater intrusion into rice paddy was a direct result of sea-level rise. Raising shrimp, a lucrative cash crop that makes use of salinity, offered a clear and simple solution. "It's so much more complicated," Thomas said. "You have upstream water diversions. You have overextraction of groundwater, which is causing the land to subside. You have soil compaction, because the sediments aren't being replenished. And yet all of those other non-climate factors get erased or subsumed under this big umbrella of climate change."

"From the very outset, if you're saying the issue is a problem of climate change, when it's really an issue of the long-term accumulation of decisions around landscape transformation through water management infrastructure and land use practices, then a certain set of responses not only comes into view but becomes the only logical option," Thomas told me. In this way, foreign agencies tailor the problem to the solution they offer, whether it's shrimp or embankments.

~

Among the CEIP's first priorities is repairing Gabura Island, otherwise known as Polder 15, which lies just south of Pratapnagar. It's probably the most precarious place I've ever been and an unfortunate archetype for the coastal conundrum: a small, nearly square island surrounded by bending rivers and a wide canal. Some 40,000 people live there, but hundreds have fled in response to lingering floods from cyclones Amphan and Yaas.

When I visited, Massul Rana, an engineer at the Shyamnagar office of the Water Development Board, told me that contractors were about to bid for CEIP funds to rehabilitate the embankments along the island's full 18-mile circumference. It's estimated to cost around US$107.5 million, but overruns are expected.

Rana explained that Gabura is a special place. So close to the bay, small, and utterly dependent on embankments, it has dealt with constant inundation and the negative effects of shrimp aquaculture. The Water

Board will test out its new designs there. The office's wooden tables were strewn with maps of the polder and colored elevation profiles of its rivers. The plan is to raise the 14-foot-tall Dutch-built walls to 18 feet and dump concrete blocks and sandbags to reinforce a critically vulnerable crook at the island's northeast corner. Like Japanese engineers, the Water Board experts have realized that more shallow slopes better handle waves, and so the new embankments will be wider and more gradual and take up more space.

To avoid waterlogging, the board plans for five sluice gates that will open wider than the old ones. To avoid conflict over shrimp farming, there will be eleven inlets intended specifically to bring saline water into designated shrimp-farming zones. To address subsidence, Rana said, "As new chars come up in the river, we'll suck sand from them and put it in the polder." The community will keep extra concrete blocks on hand for repairs, and the board will appoint a special monitoring committee. The plan does not include funds for future maintenance, however. This is typical of projects reliant on foreign aid: The World Bank covers the cost of construction and leaves repairs to the community, which, from the looks of things in Gabura, is a tall order.

I asked Rana whether the design accounts for sea-level rise.

It's calculated for fifteen years, he answered, and it should only take three years to complete.

Fifteen years. Part of me wonders whether anyone truly believes this plan will work. Even the Dutch have adopted a new approach of pulling back to allow space for their rivers at home. What's the end game to shoring up 139 polders against a threat greater than the one they've already failed to meet? Perhaps, as with the L2 seawalls in Tohoku, it's about buying time for people to flee. In this case, though, they're not evacuating from a single wave that will retreat; they're migrating. Forever. It's already happening.

Who Are We Adapting For?

Many of the people I met in villages across the southwest were left behind after their fathers, sons, and daughters went to work in garment and brick factories, drive rickshaws in Dhaka, or answer phones for Qatari businesses. Expectations are that environmental disasters will chase 35 million more coastal Bangladeshis from their homes by 2050.[68] At that time, there will be more people living in urban than rural areas. This presents a drastic cultural change for an unprepared country.

In search of better options, 400,000 people move to Dhaka each year, only to find themselves strangers in their own land. There's no blueprint for overcoming this hurdle, no engineering marvel that can solve it, no amount of concrete that can make it go away. Migrants settle in helter-skelter clusters of tin shacks such as Korail, the city's largest slum, where I meet a man named Rafik when I return to the big city on my way back home. He is forty-seven years old, a father and husband, and he works as a security guard in a commercial building. He hasn't found wealth as he hoped, though. Rafik says the twenty-six years since he migrated here have jaded his view of city life. "Once, I encouraged my family to come to Dhaka, but no more. Dhaka is not a good place for migrants now," he tells me.

For poor migrants, the city is mean and polluted. It offers low wages and the constant threats of being evicted or burned alive in a shanty community. It can do little for refugees rattled by fear, loss, and constant uncertainty. In this concrete jungle, they struggle amid a culture that stigmatizes all mental illness—from schizophrenia to posttraumatic stress disorder—under a single word: *pagal*.

Regardless, this remaking of the delta's human landscape presses on. From her research, Kasia Paprocki, a geographer at the London School of Economics and Political Science, said it's all part of the plan. Not a new plan, but a continuation of one that's been playing out here for centuries. First, she told me, Bangladesh was the world's "basket case." Then it was a laboratory for developmental aid projects. (I think of the crumbling brick structures I found in Pratapnagar.) Now, according to Paprocki, the delta provides a place to experiment with coastal adaptation. "That sounds nice and all, unless you recognize that it makes lab rats of Bangladeshi people," she said.

Shrimp aquaculture is a good example of how climate change has made development seem inevitable. It has been presented as an adaptation to rising seas. It also paved a way for rural people to enter the more lucrative industrial export economy, but there, the traders and processing plant owners had the most to gain. And what of those who *wanted* to farm? They were saddled with the environmental harms and social inequities that Karunamoyee Sardar died resisting. For them, "the alternative economic futures are not promising," Paprocki said. "Working in a sweatshop or in one of the shrimp de-heading facilities—this is not the beautiful, vibrant, agrarian future that people in Polder 22 had imagined for themselves."

Like the embankments, intensive shrimp aquaculture passes along risk. Of all its faults, this might be its most maladaptive. In Polder 22, I found the opposite: someone risking it all right now so that future generations would be spared the burden. It reminded me of a question Dilip Datta, the professor at Khulna University, posed the first time we met.

"Who are we adapting for?" he had asked, waving his hands through the cigarette smoke in his office. "Whose security do we want to ensure? If it's the third generation, then we should use nature-based solutions like TRM. If only this generation, then we should build embankments. If we manage the delta well, the problem will ease over time. If we put up embankments, the problem will be here for hundreds of years."

Datta spoke freely, and at times his perspective seemed callused. I got the sense it was the result of having been ignored by the establishment for so long that perhaps he doubted anyone was listening. The devastation caused by cyclone and monsoon flood was nature taking matters into her own unsympathetic hands, he said. "Humans have failed to keep the natural system in place in the Bengal Delta. Nature always favors the future. It is only concerned with future generations. In this way, nature always succeeds."

I had carried this thought with me as I toured the rice paddies of Polder 22 and walked in Karuna's footsteps. At first, I was struck by the abject poverty. Ramshackle houses with mud floors. Women inside crouched over charcoal stoves. Men retelling the tale of the last wave of destruction and loss. Garments hanging from clotheslines in yards fenced by World Bank–blue fishing nets. A scrawny cow ruminating in a corner. But then there were the children.

They were usually barefoot and shirtless, with curly black hair and tough little fingers that toyed absently with some tool or toy while they sized me up. Inquisitive brown eyes. They ran from the woods to the fields, stopping in at houses along the way to call on each other. They followed close behind me, skipping at my heels and lingering when I talked with their parents. And many times their parents were nowhere nearby.

"Is this your child?," I would ask a farmer. "No," he would reply as he pulled one out of a ditch or scolded them for harassing some baying goat. Here, as in all the villages I saw, the responsibility for the next generation transferred to any adult within their immediate orbit. In this way, the kids ran more freely than any I had seen at home. In the course of the coming adaptation, this is something we cannot afford to lose—in Bangladesh or anywhere.

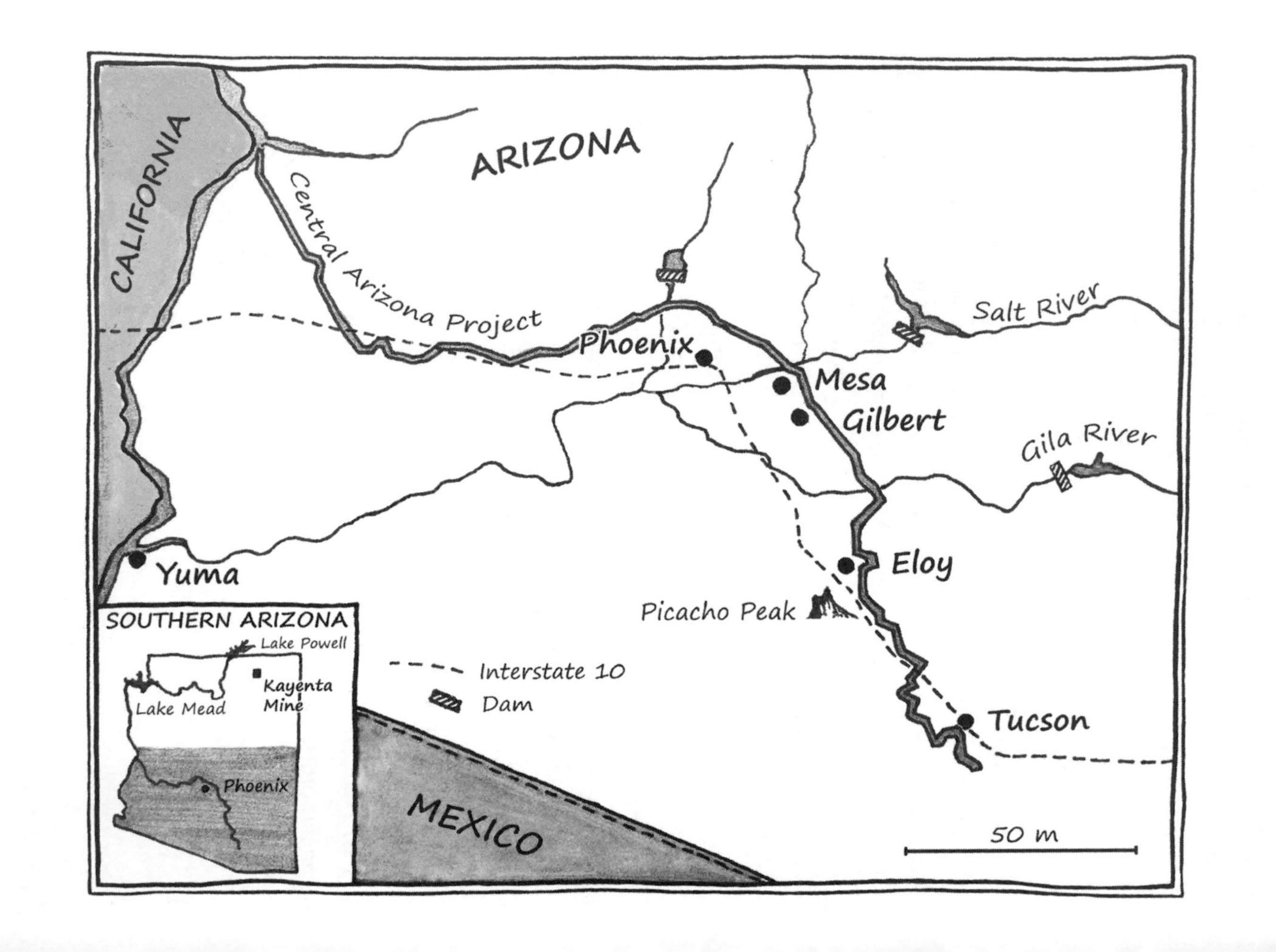
ARIZONA
CALIFORNIA
Central Arizona Project
Phoenix
Mesa
Gilbert
Salt River
Gila River
Yuma
Eloy
Picacho Peak
Tucson
Interstate 10
Dam
MEXICO
50 m
SOUTHERN ARIZONA
Lake Powell
Kayenta Mine
Lake Mead
Phoenix

Part III

The Audacity of Desert Living

Central Arizona

"We got thirty years out of the Central Arizona Project canal, and now we're back to square one," says Jace Miller, a farmer in the desert south of Phoenix. "It was kind of useless to spend the billions on infrastructure, improvements and government funding to use that water." After Arizonans nearly sucked the underground aquifers dry in the mid-1900s, the massive canal was built to bring Colorado River water to farms and cities, including Phoenix and Tucson. But now, due to shortages caused by overuse and climate change–driven drought, farmers like Miller are the first to be cut off from the river and have returned to pumping groundwater with no idea of how long it will last. "They didn't prepare for the worst," Miller says of the project's planners. "They thought it would never get bad."

"Ever downwards water flows,
But mirrors lofty mountains;
How fitting that our heart also
Be humble, but reflect high aims."
—Empress Dowager Shōken[1]

At quarter-past 10 in the morning on a blistering late-summer day, Cole Cannon surfs the desert sand. He takes a sideways stance, spreads his arms wide for balance, and runs the path of some future wave. In his mind, he's riding in the pocket, cruising ahead as the lip curls over and foam boils at his heels. Back on shore, drinks are poured and the Beach Boys pluck a low note. Everyone is happy. Life is good. In reality, Cannon is galloping around a dirt hole way out on the eastern edge of Phoenix. I squint and strain to see what he sees, but the sun is glaring, and I know from too-few childhood road trips that the nearest beach is eight hours west. We're on the fringe of Mesa, between a road named Power and the open, arid desert, so far beyond the grind of the city that it almost feels like part of something else. We are standing on the site of Arizona's newest beachfront.

Cannon is a lawyer and father of six who recently relocated from Utah with an ambitious dream. Months ago, this 40-acre plot had been a hay farm owned by remnants of farming families who worked here for several generations. In less than two years, if all goes according to plan, it will be a $25-million surf resort. Only three other such water parks exist in the United States, and despite having never built one himself, Cannon is hellbent on his being the gnarliest. Its central feature is a 2-acre lagoon. At one end, a series of steel flaps will push water into a programmable variety of 6-foot waves crashing every fourteen seconds. There will also be a 20-foot-deep standing pool that will churn a continuous, stationary wave where surfers can hone their technique. At the park's far end, there will be a sandy beach, a bar and grill, full-service cabanas, a surf shop, a four-star hotel, a gym, and a salon.

A few months before I arrive, a couple hundred people gathered for a groundbreaking ceremony in what was then still a farm hip-deep in green hay. Moms and kids, politicians and professional surfers came from as far as California to christen Cannon Beach. It was said many times that they were all super stoked to be there. Cannon is licensed to practice law in California but chafed against its regulations. "It's a terrible place to do business," he tells me. So he came to Mesa, where he found a city council as antagonistic toward red tape as he is. Mesa is the state's third-largest city, with half a million residents, and is growing fast. As mayor John Giles proclaimed during the 2019 groundbreaking celebration, it's been rooting for this beach from the start.

Cannon is comfortably charming in a crisp white t-shirt and casually expensive black jeans. He wears dark designer sunglasses and pulls his auburn hair into a tight man-bun. Over the course of our conversation, he mentions his kids more times than I can track, and when he sets off pretending to surf, he points out where the lines will form when the park opens, where the waves will break, and where he will catch up with his son, who will be nearby on a kid-sized wave. He's creating an alternative to TV and videogames to entertain his children, he says. "My wife won't surf. So I'm putting in a big salon, spa, massage, and Botox. You know, all the things." He points to a printout of the project's blueprint. "My wife might go here," he says, indicating boxes labeled *retail*, "and I'll chuck my kids in the splash pad and go rip on the waves. Then we'll all meet up for gelato. That's a good Saturday!"

Even I—pasty white and surfing ignorant—am helpless against his optimism. When I first heard of Cannon Beach, I was inclined to resent it. I grew up in Arizona, went to college here, met my wife here, and have struggled to view projects like this as anything but shameless flaunting of wasteful consumption. Preppy golf courses, ostentatious fountains, fun-filled wave parks—Bah humbug! And yet, cocking my head to see what Cannon sees in his dirt pit, I might be softening. It's obvious why he chose this spot. Last year, some 58,000 people moved to Maricopa County, the seat of Phoenix. Three miles up the road, hard hat–clad laborers are slapping together 20,000 new homes in one of many developments that will house them. Cannon wants to cash in on the boom and maybe bring some joy to this parched suburban sandscape; can you blame him?

There is, of course, one glaring question. When I ask it, he answers nonchalantly. Despite recent news of a terrible drought settling on the West, of alarmingly low water levels in stored reserves, emptying aquifers, rural communities being cut off from municipal supplies, and fish baking in dry riverbeds, he tells me water is no problem at all. By some wizardry of math, and depending on the rate of evaporation off his lagoon's surface—which would be wider than and almost as long as a football field—as well as the drops lost from crashing waves and dripping swimsuits, he estimates that he needs only about 50 gallons of water per surfer. That, he promises, would be a water savings compared to the farm that had been here until a few months ago. To fill the pools, he purchased an adjacent plot with a 750-foot-deep groundwater well, and when he plumbed it, he hit water 130 feet down. "So, from our perspective," he beams, "we actually don't have a water shortage."

Regardless of how it might look beachside in Mesa, there is a crisis unfolding somewhere. City residents won't see their taps run dry any time soon, but some Arizonans are losing access to water. Family businesses are folding, towns face economic hardship, futures are unraveling, and a way of life is fading. The reason this is so easy to overlook is that it's happening almost exclusively on farms like the one that used to be Cannon Beach. And by many accounts, for many reasons, that's for the best.

Arizona is in the twenty-seventh consecutive year of a drought that has not hit bottom. Extreme heat and wildfire ravage the region. The once mighty Colorado River's flow has declined by nearly 20 percent in the last two decades, as dismal snowpack in its Rocky Mountain headwaters, evaporation off its surface, and overuse have strangled the West's lifeline. Water stored in the river's reservoirs, lakes Mead and Powell, has fallen to record lows. Seven US states, northern Mexico, and entire ecosystems depend on that water. There isn't enough for everyone it's promised to, and there probably never was. Something has to give.

By the numbers, farmers are to blame. They use three quarters of the region's water, often to grow things we don't eat. Some argue that if we did away with irrigated farmland, we could go on living, building, golfing, and, apparently, surfing in the desert forever. Sure, there are better uses for the state's precious water than irrigating low-value crops on marginal ground, but from where I stand, the notion that swapping a productive farm for a surf park qualifies as adaptive planning shows how little we've learned about our place in this landscape and how susceptible we are to repeating past mistakes.

In settling the Phoenix valley, Anglo farmers set the tone for a century of unbridled growth. They expanded until the local Gila and Salt rivers could no longer sustain them. Then they built dams to power pumps that pulled groundwater from great depths. When that wasn't enough, they turned their sights to the Colorado River. They hashed out rules for sharing its water, and, finally, they carved a grand canal—the largest public works project of its time—to the heart of the desert. With this final infrastructural triumph, Arizonans celebrated their control over nature, but really, they had kicked the can on facing the precarity of their existence. We now stand where the can landed, and the stakes have never been higher.

Although the farmers may not have realized it at the time, by watering this desert they set the stage for their own demise. Just as they built upon the ruins of an ancient people's canals, modern cities have appropriated the farmers' infrastructure. Computer chips and residential construction are the new economic heavyweights; houses rise in tightly packed rows like the cotton they've replaced. But as cities continue to grow in the face of plainly visible warnings, I can't help but wonder what makes them so sure they won't end up like those who came before.

Following on the trail of shortsighted adaptations across the planet, I've seen how overconfidence in technology and the desire for profit over human wellbeing can leave people vulnerable to a threat they hadn't imagined. Now, the signs lead me home to Arizona. Here, greed and an undaunted sense of American exceptionalism have fueled a history of life beyond natural limits—and it continues. To understand what happens when the bubble bursts and to see where the trail leads next, I leave Mesa's carefree developments and drive south to a part of the state most Arizonans hardly give a glance. In a rural community where the crisis is already claiming victims, I revisit the history of an adaptation to aridity that many Arizonans—me included—have forgotten or never known. It's the story of a solution that facilitated growth for growth's sake while externalizing its costs onto the fringes of society, and it has brought us to the brink of disaster.

Introductions

It's about quitting time on a Wednesday, and I'm 35 miles south of Phoenix when I pull off Interstate 10. I follow the road past a Dollar General, skirting an eighteen-hole golf course and the 4,600 tile-capped homes of a luxury active adult community. I roll past empty lots awaiting new houses and breeze beyond acre after acre of unplanted farm fields. This is Eloy, and for someone making the drive between Phoenix and Tucson, it roughly marks the halfway point. That's about all most people know of the small town.

Eloy has been called that since 1878, when train operators began scribbling it on railway timetables as shorthand for East Line of Yuma. Locals had once lobbied to christen their farming town Cotton City, in the aspirational vein of places like Hygiene, Colorado, and Concrete, Washington, but *Eloy* stuck. While Phoenix to the north and Tucson to the south filled out like favored sons, it has remained someplace most of us drift past unknowingly. If you happen to be driving by, look out for the ominously dark spire of Picacho Peak, which marks the site of the westernmost battle of the Civil War, a Confederate victory. Even then, though, the Yankees were on their way to Tucson.

I turn at last down a dirt drive that twists behind a small airstrip and park in a gravel lot. Outside my car's protective bubble, it's 111°F and cloudless. I savor the last breath of icy air conditioning before prying myself free and crossing a gravel lot to a squat building with sun-faded beer signs. Squinting against the sun, I pull open its metal doors and stumble into the Bent Prop Saloon like someone being born in reverse. Once my eyes adjust to the darkness, I scan the wooden bar. At the end, under a TV blaring some forgettable game, I find Jace Miller nursing a Bud Light.

His free hand clutches his phone. He wears faded Wranglers and a well-worn, blue button-up shirt that spills a little over a silver belt buckle. His short, sandy hair is tousled under his sweat-stained trucker's hat, and he still has a bit of a baby fat beneath a patchy blond beard. Although young, he's built like a man who's spent most his life working outside, albeit lately from the seat of his truck. August is peak harvest season, and he's been up since before dawn overseeing crews spread across some 8,000 acres of desert farmland. He looks tired.

Jace is one of those lucky people who has always known exactly what he wanted to do. He began lending a hand on his family's farm during summers and holidays at the age of nine. By junior high, he was working every day after school. When other kids were learning to tackle, he was learning to drive a tractor. When they were rushing fraternities, he was pulling double shifts baling hay. He has grown to love the long, hot days, the uncertainty, and even the acts of God. Now, at age thirty-one, he's gearing up to inherit a century-old farming business from his father and says there are only two things that could force a career change: "For me, it's bankruptcy or death. I'm a farmer," he tells me. Over the past few years, that commitment had been put to the test.

The Millers (we are not related) run their farming business from near here and grow mostly alfalfa and hay on about 1,500 acres spread across Pinal County, which encompasses the broad and dusty basin between Phoenix and Tucson. Despite being in the middle of the open desert, Pinal has emerged as Ground Zero for the West's water crisis because of hundreds of local farmers who have depended on river water as delivered by that monumental adaptation, a canal called the Central Arizona Project (CAP), since the 1980s. They now find themselves on the chopping block.

Since 1922, Arizona has had the right to use 2.8 million acre-feet of Colorado River water, which accounts for 36 percent of its total consumption.[2] (An acre-foot is enough water to cover an acre 1 foot deep, or 325,851 gallons, which is almost half the capacity of an Olympic-sized swimming pool.) The rest comes from groundwater, other rivers, and a small amount of water reclaimed from wastewater treatment plants. Central Arizona receives 1.4 million acre-feet of the Colorado's water via the CAP canal, and, historically, 400,000 acre-feet of that supply has gone to agriculture. Now, the Bureau of Reclamation, which manages the Colorado's water, needs the states to collectively reduce their use by 30 percent, or up to 4 million acre-feet, in 2023.[3]

The states, cities, tribes, and irrigation districts that supply farmers hold rights to draw from the Colorado River and stand in line in order of when they arrived. Across most of the West, farmers were some of the earliest European arrivals and so often hold senior rights among settlers. This gives them priority, or the legal power to keep irrigating in the face of crisis. Although water use can be regulated, these rights cannot be destroyed.[4] This means many Western farmers are in good shape—California's Imperial Valley, for instance, holds some of the most senior rights to the river—but things are far more complicated in central Arizona. Here, growers traded away their seniority years ago for an adaptation they thought would save them.

Two days before Miller and I bellied up to the Prop, the US Bureau of Reclamation issued a long-heralded but nonetheless flinch-inducing announcement: To avert a catastrophe, it would be slashing the amount of river water that comes here via the canal. Of the 40 million people who rely on the river, Miller and his neighbors carry the unenviable distinction of being among the first to be cut off from it. As the drought has dragged on, Pinal County farmers have already fallowed—or stopped planting—about 57,000 acres, or about 40 percent of their total acreage. Somehow, they will have to survive without river water in 2023. That's a tall order for a farm in the middle of a desert, and it signals a seismic cultural shift.

"Agriculture built this state," Miller tells me. Towns like Eloy sprang up as places for growers, ranchers, and dairymen to hang their hats after long days loading harvests onto train cars, orchestrating land deals, and playing politics. They scheduled meetings to decide how the land and water should be shared and then dictated who could attend. Back then, ranchers and cotton growers threw their weight around with a confidence that has been lost on the new generation. "No one wants to do this work anymore, especially in this state, which is sad because we have some of the best growing weather in the world," he laments.

Although it may not look like prime agricultural land to most people, Miller watches the heat dance off the desert and recognizes power. He sees short winters, rare frosts, and 300 days of free plant fuel. When the Corn Belt is knee-deep in windblown snow, he's going back for a seventh harvest.

"There's arguments that say we shouldn't be wasting water on farming in a desert environment. My rebuttal to that is, Hey, we can grow—not every crop in the United States—but most of them more efficiently and effectively because of our climate. We can grow alfalfa, cotton, Bermuda, Sudan, wheat, barley, sorghum, corn, potatoes, sugar beets. We're the winter lettuce capital of the world. Broccoli, sweet corn. The only thing we can't grow here is soybeans. Well, you can grow them—they grow badass—but when we combine them, they shatter and drop on the ground."

He spits into the narrow mouth of an empty Bud Light bottle and orders another. "If I were supreme ruler of the world, there would be no one living in this goddamn state except ranchers and farmers," he says. Today, one third of Arizona soil is put to use for growing crops or raising livestock.

Years ago, Miller had earned a full-ride scholarship to the University of Arizona, and he studied business for a time, but he returned home to help his father and grandfather with harvests at a moment when the business was struggling. His sister lives up in Scottsdale, with no interest in farming. His mother left long ago to pursue her addiction to methamphetamine. The bartender reaches her hand toward his empty, but Miller beats her to it. "I know, it's a disgusting habit," he says sheepishly, and holds onto the bottle.

Jace's girlfriend, Lexi, also works on the farm. "We wouldn't be together if she didn't," he says. They employ twenty-two others. The youngest is seventeen and the oldest is in his seventies. Some are couples. An entire nuclear family works for him. Many have been with his family for years. He feels that he owes it to them to keep the operation running, but the obligation goes even deeper. "I want to carry on our family legacy, but day to day I don't know if I'm going to have a job. It's a pride thing. I get to get up every day and do it with my grandpa, my dad, my girlfriend, and I look at all my employees; I'm worried about all them," he says. "Will I be able to keep them employed? It's discouraging. Every day I wake up with a cloud over my head that never goes away."

Soon after we first met, he and Lexi announced that they were expecting a child—it's fair to say that he's all in now. Everything he has is invested in this blistered desert. In that, he's not so unlike the great bustling metropolis to the north.

Adapt or Leave

The desert that Miller farms never beckoned. It never glinted with abundance like the coasts and the plains. It was full of pointy and poisonous things whose defensive demeanor whispered of hard times. But that didn't stop American officials from looking for ways to exploit it. Beginning in the mid-nineteenth century, parties of sweat-stained explorers journeyed hundreds of miles into the region around the Colorado River on foot and horseback. They met with Native peoples, drew maps, and measured water flows. None returned to Washington with glowing reviews of what they'd seen.

In 1857, a young Army lieutenant named Joseph Christmas Ives, who set out to determine whether the Colorado River could open a commercial artery into the West, expected that he would be among the last White people to bother visiting the Lower Colorado River. "The region last explored is, of course, altogether valueless," he wrote. "It can be approached only from the South, and after entering it, there is nothing to do but leave."[5] About a decade later, John Wesley Powell, the famously one-armed Civil War veteran, saw the problem with the lands west of the 38th meridian clearly: They were too dry to farm. Expanding into them as Congress so clearly wanted to do would require a Herculean effort to redeem the empty spaces from "their present worthless state."[6]

I was also slow to fall in love with the desert. For most of my childhood, I only wanted to grow up and get out, but I came to see the charm for the crust. The change happened one spring while I was working the opening shift at a Starbucks drive-through on the east side of Tucson, where the city relented to sand. The window through which I passed venti Frappuccinos at half-past four in the morning faced east. Between small talk with the regulars I watched the sky fade from deep black through a million shades of blue while Venus hung like water caught in a spider's web above the Rincon Mountains. The sublime morning air was cool and still. Mourning doves cooed from cactus perches, and roadrunners skittered across loose rocks. In the low light, the world was clean and limitless. But the moment was fleeting.

By May, the soft veneer of spring has all but worn off and been replaced by the tan wash of summer. Heat distorts the horizon as if seen through antique glass. Kicked-up dust lingers in the air. Critters scurry hurriedly to shady holes, barely making contact with the scorched sand. May is when the realization of summer returns, when the harsh truth of another arduous season settles in earnest. Squinting against the glaring light, anyone can see the paucity of the desert's resources, the way nothing grows too tall or too wide, the way saguaro cacti allow each other wide berth on the desert floor, the way stunted mesquite trees cling hopelessly to the banks of dry washes. They're waiting. It could be a long time.

The Sonoran Desert covers more than 100,000 square miles from the Salton Sea east to New Mexico and from Prescott, Arizona, south along the narrow peninsula of Baja California. Precipitation is erratic, averaging less than 15 inches of rain a year, most of which falls in torrential sheets during the summer monsoons, and many parts receive much less. Yuma, near where Arizona, California, and Mexico meet along the Colorado River, receives a paltry 3 inches. What most of the country would consider summer comes to the desert sometime in May and sticks around until late October. Across its lower elevations, temperatures easily reach 118°F and have been climbing higher lately.

The Sonoran is a relatively new desert. Its current extent into Arizona and its unique menagerie of life probably came together just 4,500 years ago, making it younger than many of the forests and prairies of North America.[7] Like almost everywhere else on the continent, it was once a temperate rainforest, but then mountains began to lift around it. Sometime between 15 and 30 million years ago, jagged peaks rose to 10,000 feet above sea level—high enough to block wet air drifting in from the Pacific. In the lowlands between the tall ranges south of the Colorado Plateau, trees that had evolved in wet forest gave way to alien species like the now iconic saguaro.

The main stem of the Colorado River begins high in the Rocky Mountains of Colorado. It meets the Green River in eastern Utah and then cuts through the soft sedimentary rock of the Colorado Plateau, heading southwest on its way to where it trickles feebly through northern Mexico and peters out just before reaching the Sea of Cortez. All told, it drains some 244,000 square miles. On paper, at least, every drop of it is accounted for, and very little if any of it ever finds the sea.[8]

The lower stretch of the river's 1,450-mile length was carved about 5 million years ago.[9] Only the final 350 miles pass through the Sonoran Desert, entering somewhere near the town of Needles. Unlike the millions of people who now live in it, the desert does not depend on the river. The Colorado, Green, Salt, Gila, and San Pedro rivers trail lonely riparian ribbons through the dirt, but in the vast gaps between their tendrils, flora and fauna adapted to extreme aridity. Mesquite trees bear small, water-retaining leaves. Jackrabbits sport enormous, heat-shedding ears. The Couch's spadefoot has patience.[10] This small, olive-colored toad dawdles for months below ground waiting for the splatter of summer raindrops that lure it to the surface. There, in fleeting puddles during the course of just over a week, it mates and lays eggs, its eggs become tadpoles, tadpoles grow into toadlets, and these wriggle down into their own hideaway burrows, where they too will remain until another rainy season.

So despite how it appeared to wayward travelers like Ives and Powell, the desert was not a barren landscape unsuited to life; it had just taken time evolving into something specific and uniquely bound by geographic, geologic, and climatic limits. Anything or anyone who had been here long enough to notice had adapted or gone away.

Great Expectations, Shaky Foundations

The morning after meeting Miller at the Prop, I climb out of bed and head south to see him work. The address he gave was a crossing of two rural roads north of Eloy. When I arrive, he's under a tan cowboy hat handing out the day's marching orders to a small crew. Square bales of greenish Bermuda grass are stacked high on one side of a dirt lot. Tractors, balers, and other heavy equipment idle in the shade beneath a large pole barn at the other. They have about two weeks to pull in the harvest. He instructs a crew member about taking a load of hay to Marana, where it will be distributed to local horse owners. Then we jump into his white Ford 150 Super Duty and head north.

Most days, Miller lives out of this truck. Field work is largely mechanized, and hired hands handle what machines can't. Today, Miller shoots straight lines across flat desert, fielding phone calls and checking on his fields. With one hand nudging the wheel to center, he brings a Coke can to his lips and spits into the small opening. People in Phoenix think farmers here don't grow anything they use, he says, but that's not true. Pinal County ranks in the top 2 percent of all US counties in the total value of agricultural sales and the top 1 percent in sales of cotton, milk, and cattle. It produces about 40 percent of the milk and 45 percent of the beef sold in Arizona.[11] Much of what Miller harvests goes to support these local dairies and ranchers. A good portion also ships overseas to buyers in the Middle East. This highly controversial arrangement has helped Pinal growers rise to 39th among the nation's 3,143 counties in value of products sold while supporting all the local shops, distributors, and seasonal workers who rely on farm sales.

There are more than 900 farms in Pinal, and they are primarily family-run operations like the Millers'. Many lease rather than own their acres, and they have lately watched their yearly expenses rise by almost a quarter as their incomes fell by a third.[12] "Problem is every year our rent goes up and our water allocation goes down," Miller says. He let 300 acres lay fallow last year and will have to dry out more next season. Each unplanted acre represents lost income. Economists at the University of Arizona have estimated that a loss of 300,000 acre-feet of delivered water could lose the county $104 million in sales and almost 500 jobs.[13] The cascading impacts that occur as farmers call it quits entirely could be much worse.

"We're doing something we're not supposed to be doing. We're farming in the desert," Miller tells me. Still, when the COVID-19 pandemic crippled supply lines and emptied grocery store shelves across the state, there were suddenly calls for food production within reach of so many people. "We're the bottom of the food chain, but we're a necessary evil."

It wasn't always that way. Miller sweeps his spit can north in the general direction of Phoenix. "That was all ag," he says. "It was designed originally to be an ag epicenter with development on the outside, in the kind of rocky desert where it wasn't good arable farm ground." Now, a squat gray skyline dominates the horizon. The view is only the latest incarnation of a transition that began over a century ago. "No one ever thought the valley was going to do what it did," he says, his eyes lost in the distance.

~

For centuries before the Millers arrived, an ancient people whom historians call the Hohokam had watered this valley with an ingenious system of canals. When they abandoned it around 1450 ce, for reasons we're still making sense of, they left behind a blueprint for survival. Four hundred years later, European settlers established their own irrigation works directly on top of the Hohokam system. Like most migrants, they were a placeless people. They had no connection to the land and no history to teach them how to live on it, so they railed against it.

Early White settlers had little experience with irrigated agriculture in arid locales. Their towns boomed and busted with the desert's erratic seasonal rainfall, expanding in wet times as if water would never again be a limiting factor.[14] They plowed up fragile oases, riparian valleys, and riversides that anchored biodiversity. Then, during dry times, farms failed and townsfolk jumped ship.

They had come into a country that was already home to the Ak-Chin, Maricopa, Akimel O'odham, Tohono O'odham, Yavapai, Apache, and Hopi, many of whom trace their lineage to the Hohokam.[15] These people had been cultivating cotton, beans, and corn, and their skill had rendered the area into a breadbasket. Their farms supplied food and fiber for Spanish missionaries, American ranchers, and hard-rock miners long before Phoenix was settled, but that didn't account for much during dry times. In 1887, White farmers in Florence set the tone for a century of settler–Native relations when they diverted the flow of the Gila River into their own fields. The move strangled the tribes' water supply and caused nearly half a century of famine on tribal lands.[16]

And so it went. Easterners flooded westward, plowed up the desert for farms, opened businesses, and built schools. They cut roads and dug wells to grow crops they were accustomed to growing, namely cotton and hay. They stored their perishable food in wooden cabinets that kept cool through the evapotranspiration of water that dripped onto them from sacks hung overhead. On particularly hot nights, they slept on their porches and listened to yawping coyotes across a land so dark it seemed to slide right into the Milky Way. Expectations for the desert outpost were high, but on maps it remained a lowly dust-ridden territory, utterly dependent on the Salt and Gila rivers. Even with the help of the Hohokam canals, these were capricious waterways. Some years they flooded, washing away bridges. Sometimes they stopped flowing altogether. Such uncertainty was a fundamental truth of the West, but the pioneers resented it.

Enriched by a string of wet years near the end of the nineteenth century, the settlers expanded their farms across some 120,000 aces, about half of Arizona's total farmland at the time. Then disaster struck. The drought of 1897 was the worst anyone had ever witnessed. Rivers and canals dried up, livestock collapsed, and orchards withered. Farmers fallowed a third of their farmland, and armed men rode ditches on horseback looking to shoot water thieves. Phoenix began to hollow out, and the hope of a verdant Arizona appeared to be toast. To many, the Arizona experiment was over, but Alexander Chandler saw an opportunity.

Chandler was a Canadian veterinarian who arrived in 1887 and, like all successful ad men, recognized that the real money was to be made selling the dream.[17] While drought crippled his neighbors, he bought up foreclosed plots at a steal, then joined a cadre of other wealthy boosters who lobbied for federal support to build a dam on the Salt River, which pours out of the White Mountains east of Phoenix. The best estimates at the time figured that damming the Salt would allow valley settlements to expand to about 180,000 acres, but when the boosters went to Washington with hat in hand, they pegged its potential at a staggering million irrigated acres.

"No greater misfortune could befall the Valley than a mistaken attempt to irrigate a considerably larger area than the increased water supply would warrant," warned agriculturalist Alfred McClatchie in 1902. At this point in the book, it's shouldn't come as a surprise that he was ignored.

That year, Congress passed the National Reclamation Act, also known as the Newlands Act, which funded dams, canals, and pipelines from the Columbia Plateau to the Imperial Valley. Between 1910 and 1930, the amount of farmland in Western states irrigated with reclamation water exploded from 400,000 acres to 3 million acres.[18] Three years after the act's passage, the forerunner to the US Bureau of Reclamation hired Italian stonemasons to cut and place 350,000 cubic yards of rocks for a 284-foot-high plug on the Salt River. The Roosevelt Dam was christened as the tallest masonry dam in the world. For Chandler, it was the key to realizing his investment.

The Roosevelt Dam stored water for dry times, but its greatest contribution was its electricity, a third of which was used to pump groundwater for homes and farms across the valley. Hydropower brought the glow of modernity to the inky desert and ushered in a new era of modernism. This was the genesis of the ethos that would soon sweep along the coasts of Tohoku and fill the channels of the Ganges Delta. Engineering had proven that science and technology could control nature, remaking a terribly inhospitable land to humankind's benefit.

Under this high modernistic thought, climatic and geographic realities that had long posed insurmountable obstacles to growth appeared as mere snags in the story of Arizona's predestined rise. It was a hell of a dangerous precedent. For now, though, things in Arizona were going as planned. Statehood followed closely on Valentine's Day of 1912, and Chandler promoted his new town as "the Pasadena of the Salt River Valley." He sold his foreclosed tracts for $100 an acre, which at $90 profit per acre helped make him a millionaire.

King Cotton

Little grinds Miller's gears like the city slickers who come down from Phoenix to tell him how to farm. He's all for innovation and does what he can to work more efficiently, but at the end of the day, some decisions are out of his control. Climate change is only one example. Like many of the farmers I've spoken with across the country, he acknowledges it from a distance: "We affect the environment, but do we change it enough to affect life on this planet? I'm not so sure," he tells me. Either way, Miller is all for growing his crops as efficiently as possible. After all, what's "better for the farmer's bottom line is also better for everyone in the grand scheme."

The current water crisis has put his crops under a microscope, and public scrutiny has focused on two in particular: cotton and hay. Both are heavy water users; growing an acre of cotton in Arizona requires as much as 5 acre-feet, and alfalfa can guzzle 6. By contrast, wheat requires just over 3 acre-feet. In recent years, an acre of desert soil may produce 1,300 pounds of cotton and 8 tons of alfalfa.

Why don't desert farmers grow something else? is a common refrain. But the history of how cotton, in particular, came to be a mainstay reminds me of what I saw in Tohoku, Bangladesh, and other resource colonies.

The Millers came to Gilbert in 1919. Lured by boosters like Chandler, they left behind a farm in Missouri where they had grown corn for generations before. Gilbert was then a small cluster of homes and shops beside a railroad depot southeast of Phoenix, and they discovered that central Arizona was not a half bad place to grow. A wide basin streaked with fans of rich alluvial minerals that weathered out of crisscrossing mountain ranges, its soil is suited to many crops. Within a decade, the Millers and the other farmers around Gilbert were cutting six harvests of hay annually and shipping out more bales than any other depot in the United States.[19]

Hay, often in the form of the forage crop alfalfa, was like diesel. Mechanization didn't come to the Millers' farm until the mid-1930s, so alfalfa powered the mules that broke the land, fattened cattle and dairy cows, and fed horses for the rodeo. It sold for less than cotton but was common in rotation because, as a legume, it returned essential nutrients to the soil.

Cotton dates back at least seventeen centuries in the Sonoran Desert.[20] It came up from Central America in the first millennium CE, carried along with the practice of mingling corn, beans, and squash, and it did well in the region's calcium-rich, sandy loam. It needs 200 days to mature. Arizona offers 265. It doesn't like rain, which causes rot, and so prefers canal-dependent irrigation. Native people grew a variety of high-quality cotton with long, silky fibers. When Spanish missionaries passed through Native communities in what is now Arizona, they noted that after a harvest "more remains in the fields than is to be had for a harvest here in Sonora." However, Native cotton production collapsed in the late 1800s, partly because settler farmers inundated the market with a cheaper variety of cotton and partly because settlers cut off the water.

In 1909, farmers across the state harvested 27,244 acres of hay and just 19 acres of cotton. Then, around 1912, in a confluence of unrelated events, scientists at an Arizona lab developed a long-staple variety of cotton they called Pima, car companies began using air-filled tires that relied on cotton reinforcement, boll weevils wiped out cotton production in the American South, World War I severed supply lines to cotton farms in Egypt, and finally, the Roosevelt Dam was completed. Less than twenty years later, production jumped to 149,432 acres of hay and 177,013 acres of cotton, with Maricopa eclipsing all other counties.[21]

As we drive and talk, Miller repeats a steady mantra: As long as there's water, Arizona farms can outproduce just about anywhere else in the country. And in the early days of statehood, that was true. Arizona farmers raised more cotton per acre than anywhere but Missouri and North Carolina, and by wringing 2.5 tons of hay out of each dusty acre of desert, they yielded more hay than any other US state. Back then, they did so well that they hired out-of-work Okies and went on harvesting right through the Dust Bowl.

Maricopa growers excelled at raising the crop; they had to. Desert farmers had fallen prey to King Cotton. "In a sense, cotton planting is a habitual or addictive behavior, conditioned by the possibilities, pressures, restraints, and opportunities of the cotton production agrosystem and its environs," writes Erik-Anders Shapiro in his history of cotton for the University of Arizona. "Arizona growers are generally locked into a system that almost demands the production of cotton."

Ultimately, it came down to infrastructure. The Roosevelt Dam was not a donation. To secure federal support, a coalition of growers created the Salt River Project, put up their lands as collateral, and took on a portion of the dam's $10-million price tag. They were left with a debt they could repay only by growing more and higher-value products, like cotton.[22] Tapping into the Salt River Project's network also meant laying new plumbing, which demanded capital. And by midcentury, cotton ginning companies dominated local agricultural financing. Farmers found loans for growing cotton the easiest to obtain. Meanwhile, the horse-drawn plow was becoming obsolete, and the hefty costs of mechanization led small farms to consolidate under wealthy landowners and investors. Those who lived on and worked the land ended up planting more cotton to pay rent, just like the zamindars under British rule.

This all seems asinine today: Growing cotton in a dry environment uses 20 to 40 percent more water than it does in a humid one. But the business of the time didn't care; it was colonial in nature. In Arizona, there was also an untold amount of water underground and the power, thanks to the Roosevelt Dam, to pump it. During the run-up to World War II, parts of the American West produced raw goods like cotton, beef, and ore for Eastern capitalists, serving, essentially, as internal colonies. Small-scale, family-run operations were especially susceptible to this extractive relationship—and still are. We still mine rural places for the materials needed to power urban centers while encouraging brain drain through a lack of investment in rural towns.

These days, victimizing desert farmers has fallen out of vogue. Agriculture is a thorn in the side of efforts to drag the region out of the current crisis. But I can't help drawing parallels between Western cotton growers and Tohoku's prewar rice farmers or Bengal's exploited laborers. The cotton economy took advantage of Arizona's cheap land and ample sun and filled in the desert's obvious shortcoming with technology. By the end of World War II, Arizona's farmers were pumping 1.7 million acre-feet from under their boots—more than 5.6 times what Nevada currently receives from the Colorado River. This would define generations to come.

~

In the midst of all this plowing and well drilling and boostering, something incredibly important but terribly boring happened: Men wrote laws. I bring it up here only because it underlies everything that comes next. However, I recognize that for many people, digesting Western water law is like driving circles around western Kansas in a Corolla with a broken tape deck. The fact is that decisions made by a few men in the 1920s not only paved the way for the crisis we're now living but necessitated it. And in doing so, they proved that not all bad adaptations are made of concrete; some are only words.

The laws in question, collectively called the Law of the River, are a suite of policies and agreements that govern the Colorado River. They hinge on the doctrine of prior appropriation and an interstate compact signed in 1922, which together have spawned some of the most arcane legal motherfuckery I've encountered. For young Arizona, they didn't so much set a *limit* to the amount of water the state could consume as they set a *goal* that would eventually demand construction of the CAP canal and lead to our current crisis.

Prior appropriation is the Gold Rush–era dogma that established our modern system of seniority by decreeing "First in time, first in right." Basically, pioneers from the soggy East were alarmed to find that Western rivers sometimes ran dry. Their responsive capitalist instinct was to establish a pecking order that determined who would suffer in dire times, based on who got there first. Because the doctrine respects precedence above all else, it encourages water users to take all they can—to use water in order to claim water—and thereby places little value on thriftiness. This is the basis for the schism between how agricultural and urban users still view efficiency: City dwellers define it as using less water; farmers see it as growing more with all the water they have rights to.

Western states always had their eyes on the Colorado River, and when the Reclamation Act set them on a growth bender in the early twentieth century, the doctrine of prior appropriation made sure there would be a race to secure its water. Like wolves to a kill, states that developed slowest would starve as the others gobbled up the rights. California was quickest out of the gate. Nevada, Utah, New Mexico, Colorado, and Wyoming struggled to keep up, and Arizona, whose population was centered so far from the river, risked being left in the dust.

The compact arose as a reaction to this race. It was a problematic adaptation to a fraught legal environment. In 1922, representatives from the seven states and the federal government sat down with the goal of negotiating limits on use of the river that would allow each to develop at its own pace. A fine idea, until it became a fiasco. The men—all White and wealthy elites—bickered over everything, beginning with how much water was in the river, because the system is huge, spread across vast and unwelcoming terrain from the Rocky Mountains to the Sea of Cortez, and because the negotiators began with limited information.

Eugene Clyde LaRue, a US Geological Survey engineer, worried about the lack of reliable river flow data. He thought there were too few gauges, too short a period of measurement, and too little understanding about past droughts to make an accurate estimate. His early math put the Colorado's flow at 16.2 million acre-feet, which, he acknowledged, "would not be enough water to meet the aspirations of the seven states."[23] A long-held myth of the compact holds that the negotiators were merely naive when they ignored LaRue's warning and set the region on its current crash course. In this telling, they unwittingly based their assumptions on a period of unusually wet years when the river ran abnormally high. The not-so-lenient truth, however, is that politicians succumbed to ambition and greed with flagrant disregard for the truth.

"That all this should have been confusing to them is expected," write Eric Kuhn and John Fleck in their retelling, *Science Be Dammed.* "But their sin, which became the original sin of Colorado River Basin management, was a lack of humility in the face of their ignorance." Had they taken the available science seriously, "they almost certainly would have concluded that the Colorado River had less water than the common assumptions underpinning their race to the develop the river." Instead, they chose to "selectively use the available science as a tool to sell their projects and vision for the river's future to Congress and the general public. This approach used by the compact commissioners set a precedent that would continue for decades and color most of the major policy decisions on the river in a way that still wields undue influence today."

The negotiators knew any future infrastructure project on the Colorado River system would have to pay for itself. And so they fudged the data so the river appeared to hold enough economic potential from the eventual sale of its water and power to cover the cost of the canals and dams they wanted to build. After lengthy negotiations, the group concluded that the river could provide at least 17.3 million acre-feet and possibly even 22 million. They signed an agreement that legally committed 16.5 million acre-feet to the seven states, 1.5 million acre-feet to Mexico, and an additional 1.1 million-acre feet to the driest states in times of surplus. Arizona, in which about 330,000 people lived, would be given rights to 2.8 million acre-feet. There was not a drop to spare.

Did these men think of their work as adaptation? I doubt it. The word doesn't appear in the compact, and although *The Origin of the Species* had been published sixty-three years earlier, adapting was still something only birds and plants did. Still, they were altering behavior in response to environmental circumstances. By deliberately overestimating the river's flow and committing to the delivery of water that didn't exist, however, the negotiators adapted us all into a corner. They impelled unsustainable growth while robbing subsequent generations of the flexibility to adjust amid unseen changes.

Today, more than 7 million people live in Arizona, with more on the way. Meanwhile, the Colorado carries little more than 10.5 million acre-feet, and, thanks to continued overuse and climate change–driven drought, that number is falling.[24] In effect, we are left with two Laws of the River: the 1922 compact and the physical code of hydrology. It's a testament to the languishing power of the past that each now *feels* as immutable as the other.

A Glimpse of What's to Come

Jace and I are cruising. Rows of green alfalfa, tufts of white cotton, and patches of weed-infested dirt stream by through the bug-streaked window. Power lines unspool overhead. They lead us along a slender line of fresh asphalt through land that stretches out toward nothingness. At times, Miller rants about the underappreciated role of the American farmer or the lack of work ethic among his generation. I search the flatness for signs of life. Well off the road in the sun-bleached distance, occasional palm trees divulge farm houses. Most are lonely brick structures flanked by rusting equipment and browsing horses. The scene appears a far cry from bustling Pinal County of the mid-twentieth century, when growers went on a tear without realizing they were working against their own best interests.

Back then, land was cheap, the sun was cheaper, and as long as they had water they could fatten cattle and grow more crops more times a year than just about anywhere else on the planet. The technological boom of World War II catapulted cotton into an industrial juggernaut, its grip on Arizona's agriculture tightening with new machines, fertilizers, and churning factories. Planting peaked in 1953, when farmers sowed across 695,000 acres, mostly in Maricopa.

Visiting Arizona's cotton and alfalfa farms, farmer-cum-environmentalist Wendell Berry saw what decades of cotton monocropping had wrought in the desert. "This is modern industrial farming in its purest form: enormous, costly fields, dependent for their productivity on large machines, fossil fuels, chemical fertilizers, insecticides, and herbicides," he wrote in 1979. "Precarious as these dependencies are anywhere, in Arizona (as in the Southwest generally) another even more critical dependency is added: fossil water."[25]

The cotton industry depended on the aquifers beneath Arizona's crust, which the state's residents treated as infinite sources. While state leaders fought over what to do with the Colorado River, they drew three quarters of their water from below, but groundwater proved to be a treacherous elixir.

Picacho Peak, that ominous spire that casts an evening shadow across Interstate 10, is the remnant of a volcano that formed some 35 million years ago.[26] Characteristic geology of the state's lower half, it's part of a basin and range where shattering faults birthed jagged mountains and opened wide valleys. The lowlands are filled with rubble that washed out from above, mostly granite weathered to sand and gravel. These basins might have funneled their waters into new channels that joined the Salt, Verde, or Gila rivers if not for impenetrable underground rock walls that resisted their flow. Instead, water backed up beneath the surface, seeping into fractures and permeable substrates to form underground lakes and pools. On average, these aquifers fill at a rate of just 2 percent each year, and the last time the continent was wet enough to top them off was during the last Ice Age, over 10,000 years ago.

Arizona's first real estate boom had erupted in the 1920s. It built like monsoon clouds over a late June day and burst with a deluge of housing development spurred by generous subsidies and the suburb-loving automotive industry. From 1940 to 1960, Phoenix's population ballooned from 65,000 to 439,000, eclipsing for the first time what the valley had sustained with the Hohokam canals.[27] Meanwhile, Tucson constructed more than 41,000 new single-family homes, many for the snowbirds, temporary residents accustomed to colder and wetter environs who consumed water and electricity in the piggish habits of their greener homelands.[28]

The fact that reservoirs like the one behind the impressive Coolidge Dam, which was built on the Gila River soon after completion of the Roosevelt Dam, had never received enough runoff water to fill might have been cause for concern about the desert's rate of replenishment. Instead, on farms, in mines, at lumber mills, and across the bustling new cities—but mostly on farms—Arizonans pumped mercilessly from their wells. At midcentury, about 70 percent of the water used in Arizona came from underground.[29] The state sucked 500 times as much water from the aquifers each year as what the wispy clouds could replace, so by the 1940s, the drying desert soil began splitting into wide cracks and dropping beneath them.[30] Thirty years later, nearly 625 square miles around Eloy had subsided, by as much as 12.5 feet in places. The water table under Tucson dropped 200 feet, causing land subsidence and the virtual disappearance of the Santa Cruz River.[31] In the mysterious, unknowable underground, the water table was collapsing.

To avert catastrophe, to survive, but mostly to grow, Arizona needed a new source of water. Technocrats had already dreamt up pipelines to the Columbia River and floated the idea of dragging freshwater in the form of icebergs along the Pacific Coast. But these were harebrained fantasies. Some in Phoenix had been calling for a canal to the Colorado since the 1920s, and this appeared to be the only way Arizona could make use of the 2.8 million acre-feet of Colorado River water it had been granted rights to under the compact. State leaders fretted over the prospect of California sucking up *their* water. The challenge was that most of the state's people had settled hundreds of miles from the river and could hardly imagine, much less afford, the infrastructure to bring the river to them. Such a canal had always been a "mad man's dream."[32]

This was a critical moment, one that I believe is not so unlike where we find ourselves today. Had those tilling and paving the desert soil paused long enough to let their eyes drink in their depleted landscape, they might have seen that it was not the lavish land they had come from and never would be. Over the previous fifty years, each new dam and reservoir had opened new farmland and brought more settlers who irrigated and pumped until they exhausted their stores and demanded the cycle be repeated. This might have been the time to take stock of certain unalienable truths, to save future generations from unending conflict and worry. Instead, Arizonans turned their attention to the Colorado River. From the brink of catastrophe, madness made sense.

Leaders set about drawing technical plans for a canal connecting the Colorado River with Phoenix and Tucson and convincing the federal government to fund a monumental public works project larger than anything the United States had ever built. But there was one sticking point: During negotiations over the compact in the 1920s, California had called on its seniority under prior appropriation and demanded assurance that in times of drought—something that by now was inevitable—its needs would be met before Arizona's. The Grand Canyon State had obstinately refused to accept Californian preeminence and remained the compact's sole holdout for decades. Now, it was backed into a corner with emptying aquifers and an anxious population. Bow to California, the Feds said, and we'll pay for your canal.

In 1968, Arizona bent a knee.

The Millers remember this time. Jace's father was in the position the younger man holds now, preparing to take over the farm amid deep uncertainty. They viewed the proposed canal as a godsend and a monument to their country's mastery of the natural world. The American farmer had played a critical role in fueling the country's victory overseas and its continued progress at home. By linking fields in central Arizona with the Colorado River, their country was making good on its responsibility to keep farmlands productive.

A canal would also untether the state from the limitations of its paltry surroundings, which would effectively obliterate the notion of a carrying capacity. If successful, Arizona would exist within a biosphere of its own making, bounded only by measureable and manageable inputs and outputs, the high modernist's dream. Fifty years of hindsight have revealed that this notion was ludicrous. Rather than realizing human domination of the land, the experiment in the desert has proven, in Berry's words, that any attempt at total control is ultimately an invitation to disorder. "And the rule seems to be that the more rigid and exclusive is the specialist's boundary, and the stricter the control within it, the more disorder rages around it."

For the farmers, the trick, then, was to stay on the inside, but for others that was never an option.

Power for Water

It's easy, with foreign eyes, to see the misery that bad adaptations have wrought in a place like Bangladesh. Walking through villages where residents rebuild from flood time and again, I recognized how centuries of miscalculation and arrogance had undercut the futures of certain people. Back home, this is not so obvious. We are raised with stories that provide the lines, and we paint within them for the rest of our lives.

My public school education of Arizona's history was heavy on the intrepid pioneer, light on the ancient Puebloan, and devoid of anything in between. There was a gaping hole between Brown people grinding corn in the dirt and White people deftly reclaiming the desert. But the European farmers who staggered into Arizona in the nineteenth century didn't fill a void. They elbowed out others whose crops had risen toward the sun for thousands of years and gradually pushed these backward people to the geographic and political fringes.

If not for Ms. Esquerra, my fourth-grade teacher who demanded we capitalize the I in Indian and thereby bestowed some importance on what had otherwise been a footnote, I might have given little thought to those on the outskirts of the settler's experiment. The story of the canal, for instance, was always (though rarely) told as a tale of human triumph over the unabiding desert. I never heard of how the tribes around me had been cornered into facilitating their own marginalization to build it. I only knew that when I turned the faucet, out came water.

Andrew Curley grew up with different stories. The son of a Navajo man who worked as an accountant at one of the country's largest defense companies and a White woman who worked for the country's largest tribal nation, he became a geographer charting tendrils of water, power, and time across the desert. While researching how coal came to dominate and then gut his people's communities, he saw that the burdens Arizona's greatest adaptive scheme had imposed on its tribes.

"We were at the core of Arizona's development and growth, and yet we have been treated like shit this entire time," Curley told me over coffee. He was on his way home from speaking at a conference on Colorado water issues. "Our water rights have been pushed to the side. Our land was decimated and destroyed. And then our economy was just suddenly abandoned once it became less convenient to continue with coal."

Midcentury settlers had rights to the water, thanks to the Colorado Compact, support from the federal government, and engineers who could build the necessary pumps and dig the canal. But the project would go nowhere without the power needed to lift water over hundreds of miles.

Energy was the limiting factor, and the state's leaders first proposed overcoming it by building two dams along the Colorado, at Bridge Canyon and Marble Canyon. These would flood portions of Grand Canyon National Park, which was something American environmentalists could not stomach. The environmental movement was gaining steam in the 1960s. Corporate interests and the politicians they controlled got a taste of this new power when David Brower's Sierra Club put the kibosh on the dams. And that might have been it, if not for a mountain of coal in the black hills of northern Arizona's red dirt expanse, on land belonging to the Hopi and Navajo.

"The interests of the people of Arizona dictate that there be no further delay in getting water into Arizona; the prompt way to get water into Arizona, the cheapest way, and the way which will have the least opposition, is to use coal," wrote Anthony W. Smith of the National Parks Association, in a brief to Congress in August 1965.[33]

Even Brower agreed. "We are not an advocate of strip mining at the present time, but something is going to be lost. You have to weigh the kind of value you are going to lose," he told Congress in 1967. "I would say still, even though the desert is fine country there, it is not a world famous desert in the sense that the Grand Canyon is a world famous canyon, and I would say that that would be the lesser of two evils." The unexceptional desert to which he referred was Black Mesa, a high and dry expanse of pinyon pine and cacti some 50 miles south of Monument Valley where coal seams lay so near the surface that people had harvested them for eons.

The tribes had little chance of rejecting the plan for their land. Navajo and Hopi economies were in the tank, their futures uncertain, and so in 1968 when the state, the Salt River Project Coalition from the Phoenix valley, and the Bureau of Reclamation offered a way out, they accepted a Faustian bargain, relenting to a massive power plant that would burn bituminous fuel from Black Mesa to pump Colorado River water into Phoenix and Tucson. The agreement the tribes signed with Peabody Western Coal Company paved the way toward mining an eventual 14 million tons of coal per year. It also allowed the extraction of local groundwater for the operation's use.

The plant's boosters had won support for the project partly by hyping the money the Navajo stood to make from their dirty resource. Although Peabody would own and operate the Navajo Generating Station, mining and selling the coal would provide about half of the Navajo Nation's annual budget and 1,000 jobs. And Black Mesa coal was fired not only locally but also at the Mohave Generating Station in Nevada, which provides power to California. Conveying Navajo coal to Nevada by rail in the conventional fashion was too expensive for Peabody, though, so mine employees mixed the stuff with water into a slurry that sloughed 273 miles northwest from the Navajo Nation to the Mohave Generating Station. And because Colorado River water was too precious a resource to be wasted transporting coal, each year hundreds of thousands of gallons of drinkable Navajo Nation groundwater were used instead.

"They were sacrificing water for coal," Curley said, in a place where, even today, a third of families lack running water and so drive miles to haul it for their homes.

Coal brought revenue and the plant brought jobs, but all came at a remarkable cost. Churning through eons of carbon locked in Black Mountain sediment, the Navajo Generating Station became one of the worst air polluters in the country. Coal ash floated over Navajo villages and schools. Veils of smog thick with nitrogen oxides—the fine particulates that result in asthma, bronchitis, heart attacks, and premature deaths—shrouded views of the Four Corners region's pristine natural landmarks, including the Grand Canyon. The aquifer beneath Navajo lands collapsed.

With every intention of stirring up images of landing craft breaking through salt spray, Curley refers to the coal industry as a "beachhead of colonialism" on Diné lands. "In this new brand of colonialism, suits and ties replace muskets and bayonets." The incursion would not only bring water to central Arizona, it would bind tribal communities to modernized, urbanized, and capital-dependent systems of wealth. Through the use of their land and resources, Diné and Hopi people would be put to the task of providing the water and power needed to make the desert comfortable for incoming Anglos while suffering the externalities.

Viewed through Curley's lens, the Colorado River as we know it today is an abstraction of a river pieced together from measurements of flows and tributaries that were taken in service of economic goals and political ambitions, not genuine interest in a living river. We know how much water flows through the Colorado and where it comes from because that knowledge was key to unlocking a valuable resource. The term "river basin" emerges in English only in the late nineteenth century as irrigators, engineers, and administrators needed to define what they were managing.

"These are all scientific understandings that are in service of political ends," Curley said. "Science is not standing outside of politics here; science is always complicit with the politics of the Colorado River."

In the 2000s, the drumbeat of climate activism finally came to the reservation. Environmentalists sought to shut down the Navajo Generating Station as part of growing distaste with coal combustion. By this time, though, despite all the troubles the plant had caused over the years, tribal communities stood hotly divided over what to do. There was no denying the links between coal firing, climate change, and problems with local water and air quality, but closing the plant would mean losing hundreds of jobs and millions of dollars of revenue in Arizona's poorest county. In the end, environmentalism prevailed, and the station closed in November 2019. Just as it had in the 1960s, mainstream environmentalism had chosen the lesser of two evils.

"It's inaccurate to say that nothing was gained," Curley says. "There were jobs, and there was revenue from the production of the energy needed for CAP, but our water and political powers have been diminished over time. The Navajo Nation was in a much more advantageous position then than they are in today. Now, Arizona has many more representatives from these conservative suburban areas that resulted from that water expansion."

"What was lost was our ability to keep Arizona in check," he told me. "If we didn't participate in that project, I wonder if it ever would have been built."

~

On May 6, 1973, at the modern site of the Havasu Springs Resort, in a canyon the Havasupai people know well from the millennia they've lived near the western outlet of the Grand Canyon, an explosive charge sent ochre sedimentary rock careening into a cavernous gorge. That day, Arizona and the Bureau of Reclamation began building a canal that now slinks along in a glinting ribbon across the scrubby desert. From above, it resembles a badly healed scar. At ground level, it looks like some anthropogenic approximation of a river: sleek and narrow, with uniform sloping concrete banks and simple curves.

The CAP averages 80 feet across at its start and narrows to 24 feet at its end. When full, its water reaches 16.5 feet deep, carried along between 3-and-half-foot-thick concrete walls that are dyed tan to blend in with the scenery. Most of the canal's 336 miles are protected behind a chain-link fence and patrolled by a private security force. Wildlife bridges allow large animals to cross it, fencing gaps allow smaller animals to enter it, and ramps provide an escape from drowning in it. Although the CAP no longer runs on Black Mountain coal, it remains the single largest user of power in Arizona, consuming 2.5 million megawatt-hours.[34]

The Bureau of Reclamation considered covering the canal to limit evaporation but quashed the idea after realizing that would have quadrupled the more than $4.4-billion price tag. So most of the time its water winds nakedly under a blazing sun. In some places, the canal zigs and zags in concert with elevation contours. In others, it disappears under mountainsides like blue thread into leather. Engineers spent twelve years blasting a 22-mile tunnel so it could travel under the Buckskin Mountains from what is now a nine-hole golf resort on the south end of Lake Havasu to wild, open desert. From there, it flows southeast, traversing wide alluvial fans, crossing dry washes, darting through the eerily desolate Cactus Plain Wilderness, and crossing and recrossing Interstate 10 on its way to Phoenix. In the Valley of the Sun, it spreads tentacles through the metro, then continues south across the Salt and Gila rivers and on down to Tucson.

The CAP was hailed as the most significant infrastructural achievement since the Hoover Dam, but locals weren't immediately smitten. Tucsonans, in particular, felt duped. When canal water finally reached them 1992, it was brown, foul tasting, and highly corrosive to home appliances.[35] The acrid water destroyed the city's plumbing, killed pets and plants, caused rashes, and sent the city into an uproar. These problems were largely the result of nasty interactions with Tucson's neglected infrastructure. The adaptation had been undertaken so hastily that there hadn't been time to update the systems it would plug into.

"That just created a firestorm of anger," recalled Molly McKasson, a teacher and former Tucson City Council member. "Quality of life had been put on the back burner, and on the front burner was just the issue of growth, that we have to just keep growing, and the faster the better. It looked as if everything had been kind of short-sighted, and clearly planning was short-sighted."

Still, the CAP uttered the climactic grunt of the push to reclaim this parched land, and despite complaints from folks like McKasson, it was astoundingly successful. Since its first water deliveries in 1985, the canal has generated trillions of dollars for Arizona's economy and has allowed desert cities to blossom.[36] Out of the gate, from 1970 to 1990, Phoenix ballooned by 91 percent, sprawling across an additional 353 square miles of desert.[37] Farmers who had nearly pumped themselves into oblivion found new hope. Buoyed by this tenuous lifeline, they went on planting cotton, alfalfa, grapefruits, and hay, raising cattle, and watching their cash receipts nearly triple in those twenty years.

The project made some smart concessions to the desert too. Under pressure from environmentalists concerned about groundwater collapse, the federal government refused funding unless the state promised to rein in overpumping, which it did by limiting or banning groundwater withdrawals in the most populated parts of Maricopa, Pinal, and Pima counties. It also established a system to recharge depleted aquifers by dumping CAP water into the ground. These were the most progressive achievements the state has ever made regarding water management—before and since.

By weaning central Arizona's cities and farmers off groundwater, it probably saved the aquifers. However, this didn't solve the underlying problem; it merely shifted the burden onto the Colorado River while creating a dangerous illusion of abundance. In the words of the Tucson writer Charles Bowden, building the CAP was "like giving a case of whiskey to an alcoholic."

Return to the Underground

I was in a clearing between empty fields out near Coolidge, another Pinal County town founded on cotton. It was eleven in the morning and already skirting 100°F, with, in an uncomfortable change of pace, 68 percent humidity, thanks to the residue of last night's storm. The land was flat and wide and separated from the sky on two sides by sawtoothed strips of distant purple mountains. I stood near the edge of a shallow, *V*-shaped ditch that ran beside a dusty farm field. It was empty except for a layer of tan dirt and some tangled tumbleweeds. A few yards from the ditch, two young men in sweat-stained work shirts clambered over a red 1950s-vintage drill rig, a hulking relic of the pre-CAP days that had once plumbed the desert soil to bring virgin land into production. Today, it was clearing some mystery grime from the bottom of a 1,000-foot well and coughing black smoke into the limitless sky.

Justin and Moe, the grime-covered men, extended the rig's mast and lowered a huge metal syringe down a narrow hole in the earth. Then they let out its metal cord until the tool hit bottom. When they reeled it back up into the light, Justin gave the syringe a stiff whack with the metal hammer, and a black amalgamation of mud, sand, and rust spewed forth. Crud. Greasy crud from 900 feet below Earth's crust. I jumped out of the way just as it erupted across the desert sand, spilling its oily translucence into the empty irrigation ditch. Then the engine coughed, and the syringe went down for more.

"It's crazy to me that water even runs under us. And clear water," Justin exclaimed, jumping down from the rig.

Moe, the younger of the two, said he'd been with this drilling company for six years and that it beat housing construction. "The drought has actually been good for our business." That's because irrigators across the valley are hurrying back to sucking from underground. That's right: Less than fifty years after the Bureau of Reclamation spent $4.4 billion to get farmers off Arizona's depleted aquifers, here they were, pumping again.

Tommy Hoover, who owned the rig, said his company worked more than 120 wells a year, mostly in Pinal, and he expected to ramp up as river shortages continued. Most of his clients hired him to deepen or clean existing wells that had been sitting idle from about the time the

CAP was finished until just recently. Pumping around here is still heavily regulated, but he had noticed the water table drop, especially outside the urban cores, beyond the reach of groundwater pumping restrictions, where enormous corporate farms were draining the aquifers beneath rural communities.

Rehabilitating water wells isn't cheap. It can run $200,000, which is a significant investment for cash-strapped farmers. "Some just give up," Hoover told me. "We had a set of farmers here who just decided to retire. We drilled a 200-foot well for them in 2014, and then they sold the farm the very next year."

The image of that black sludge spilling across the desert soil springs to mind when Miller says, begrudgingly, "The CAP was a Band-Aid. Canal water was a temporary fix."

It brought new life to his family for some merciful seasons. They expanded their acreage, bought better equipment, and experimented with new products in the hope of gaining a competitive edge. So why are they now returning to groundwater? Because the canal was a coping mechanism, not a solution—and it was never intended to save farms. The CAP had more to offer the boosters, bankers, real estate investors, and developers who saw in the farmers' dire situation an opening to ensure water for the urban state they longed to build. Agriculture would be a placeholder, temporarily putting Colorado River water to use and keeping it out of California's hands until the cities were big enough to need it.

One year before Congress signed off on the canal, William Martin, an economist at the University of Arizona, published a study directly criticizing the project.[38] Risking his professional neck, he questioned whether the canal was needed at all, and he warned that, given the infrastructure's insanely high cost, there was no way small-time farmers would be able to afford its water.[39]

The CAP was proposed at $832 million, but the final cost had ballooned to $4.4 billion.[40] Four years after it came online, a special advisory committee created by the governor concluded "that at some point most or all of the irrigation districts [which provide water to farmers] may choose or feel compelled to seek the protection of federal bankruptcy court."

In 1983, Arizona farmers had planted just 30,000 acres of Pima cotton. The highest-quality variety, Pima or ESL cotton, produces lower yields than the hardier short-fiber cotton but can fetch 50 percent more profit.[41] Five years later, after the canal was running in Maricopa, Pima cotton acreage jumped to 245,000 acres. Farmers were struggling to grow enough to pay for the liquid gold flooding their fields. They were trapped in yet another cycle of exhaustive extraction, forced to plant water-intensive cotton to pay back the price for an adaptation they thought would save them from water shortages.

Why, then, had farmers been among the CAP's most ardent supporters during negotiations with the Feds? Curious, Martin surveyed hundreds of central Arizona growers and discovered that it came down to the fact that they had no other choice. "When they feel 'their backs against the wall,' they do not question the form in which 'help' comes," he wrote, quoting farmers' responses. In parts of Pinal, some growers spent $1,100 per acre on upgrades to convey the canal's water to their fields. Their situation was so dire that it was worth the astonishing cost.

"Price doesn't matter," one cotton farmer told Martin. "The point is we need more water. It is our last chance. The price of cotton is too low."

For his effort, Martin was accused in the local newspaper of producing "anti-Arizona propaganda" and doing "a great disservice, under the guise of academic freedom."[42] He and a partner were forced out of their academic positions. Undaunted, he went beyond the canal, arguing that even some cities' laudable conservation efforts served no other purpose than to ensure enough water for someone else's continued growth.

This adaptive failure is not unique to Arizona. Farmers in Ghana, Mali, and Australia had similar experiences in the wake of new of irrigation infrastructure: The price of water invariably rises to fund the project, causing growers to stick with what they know and what fetches the highest price. Rarely is that some experimental wundercrop. In Spain's northern Navarre region, for instance, construction of the Itoiz Canal de Navarra made farming prohibitively expensive and caused towns built on family-run wineries and winter wheat and barley to consolidate under wealthy corporations that swapped traditional crops with corn for biofuel.[43] In 2020, researchers found that the canal had dented local social life and damaged ecological diversity while "leading farmers to maladaptation to climate and market variability." The inheritors of generations-old farms packed for new lives in the city.

This is a future the Miller wants desperately to avoid.

~

It took 117 years for the Gila River Indian tribes to get something like justice for the theft of their water. In 2004, after a seventy-five-year legal battle over water Anglos began siphoning off their lands in the 1880s, the tribes won rights to the single largest allotment of CAP water. Adjudication of Indian water claims is a terrifying prospect to many farmers; it threatens to pull the rug out from under all they've built, and that's what happened here. Because there's no extra water in the Colorado, the tribes' water had to come from somewhere, and it came from farmers like Miller. In exchange for the forgiveness of a debt from the canal that they could never repay, he and hundreds of other farmers surrendered their long-term water rights to the tribes and agreed to take a place at the front of the line should there ever cutbacks on river water deliveries.

Under their initial settlement, farmers could depend on canal water until 2030. But as the drought has worsened and levels at Lake Mead have fallen, the seven Colorado-dependent states have amended the agreement to save river water, in part by gradually reducing deliveries along the CAP to central Arizona's low-seniority Anglo farmers until they reach null in 2023—seven years earlier than what the farmers were expecting.[44]

Irate growers have complained that the deal keeps changing, that the cities move the goalposts. The real problem is that they became solely dependent on such a precarious lifeline without knowing how quickly things could change. Exactly one hundred years since a group of men signed a compact ensuring that their interests—predominantly agricultural—would thrive on Colorado River water, the first farmers lost access to it. And that's how it came to be that in August 2021, Miller learned that he would be inheriting a farm with no water.

"I don't want to get out, I want to farm. I find more happiness in the legacy of the farm and raising my family that way, the way I was raised," he says. But as the days bleed by and the Colorado River dwindles, his window appears to be closing. The Millers sold their original homestead in 2005 to cash in on the housing bubble and intended to lease acres around Gilbert until they found new property to buy. Too soon, however, the investors who owned that land decided to sell, and the Miller family migrated south to Eloy. By the time Miller was considering college, they were in a tough spot. The valley was developing fast, water and land were expensive, and good labor was scarce. Someone else might have gotten out, but he decided to stay on the farm.

"My dad and grandpa didn't force me to do a damn thing, to come back to the farm. But I knew if I went to college and had done something else, I couldn't come back. They would have been gone," he says with obvious frustration. There would have been no farm to come back to. "It was a changing of the guard, and I took up that torch."

Now, one of the greatest achievements in American geoengineering, the culmination of decades of increasingly audacious schemes, which allowed agriculture to expand and prosper in the desert, has dropped him one hundred years back in time.

Tens of thousands of groundwater wells dot Arizona. The US Geological Survey measures aquifer water levels from 36,471 sites.[45] From these, it can track the rise and fall of water, but aquifers are wide and complex and vary drastically. There simply aren't enough data to know how much water is left, or where it's going. "You can't manage what you don't measure," Tom Buschatzke, the director of the state's Department of Water Resources, told CNN in 2022.[46] "We do the best we can with the data and estimated data that we have, but it really begs questions about how much benefit we can really provide."

Across from Eloy's Memorial Park is a well that's been in place since 1949. Its measurements track the rise and falls of overuse and attempted recovery. Its first recorded water level started just 190 feet from the surface. By 1984, it had fallen to 415 feet. The arrival of the CAP helped to bring it up, but its latest measurement shows its water remains 282 feet down, and as farmers return to their wells, there's no way to be sure how long it will last.

"All the growers expected to have access to CAP water until 2030," said Blase Evancho, an agricultural extension agent with the University of Arizona. "And with that not being there anymore it's got a lot of people scared, a lot of people nervous about the future. That's including me. My career depends on agriculture in Central Arizona, and hopefully it sticks around." The Bureau of Reclamation has called for further reductions of 4 million acre-feet, and everyone knows where it will be felt: farming communities like Eloy, where growers hold low priority on the river. For a time, at least, people in Phoenix and Tucson will hardly notice.

At a tractor and supply store in Casa Grande, the sales manager told me that farmers' buying habits have already changed. "They used to come in and buy new gear every three or four years. Now they're holding onto the old stuff," he said. When its farms dry up, will the valley's cotton towns look like Great Plains communities that never fully recovered from the Dust Bowl?

Bill Travis, a Western water expert at the University of Colorado, supposed that the Rust Belt might be a better analog for what's happening here. In terms of employment, the cotton and dairy industries are to Pinal what auto manufacturing is to Detroit, and their loss would lead to a similar hollowing.[47] "Eventually you have abandonment," he said. "In Detroit, for instance, people literally walked away from their houses. The city walked away from services and infrastructure."

Jace hopes on-farm improvements will at least buy time. Concrete-lined ditches retain more water. High-tech monitoring helps avoid applying water where it isn't needed. Cutting back on tilling helps the soil maintain moisture. Laser leveling the land has doubled the efficiency of flood irrigation, although pouring water across a desert field is still a touchy subject. Drip irrigation would be ideal, but who has the $2,500 per acre for it? A new problem is that their wells don't pump fast enough to make use of it. The CAP delivered exceptionally high flows of about 15 cubic feet per second, which allowed farmers to flood irrigate large areas quickly. Today, many groundwater wells pump at only 3 or 4 cubic feet per second. One of Miller's wells dribbles just 1 cubic foot per second. At that rate, it doesn't matter how much water he has, he can't spread it across his field faster than it seeps into the soil, and he ends up using more in the aggregate.

Another problem that is many Pinal farmers don't own land. Like renters in a dingy apartment, tenant farmers have little incentive to shell out for improvements, especially when they know fields are likely to fill with houses. Miller would like to flee the rental trap, but the bank won't write a loan for anything less than 2,500 acres. "Anyway, it's just not worth it," he says. "You can't make enough farming to pay off the mortgage and cover equipment." I point out that the truck we've been driving around in gives the impression he's doing all right. "We have extremely high gross numbers," he says. "People have the perception that we're all millionaires, but our margins are slim. Ten-percent net profit would be great." Half the county's farms earn less than $10,000 in sales.[48]

Access to CAP water facilitated the industrialization of Arizona's farms and negated the urgency to adopt significant adaptations by conditioning growers to dependable flows. Now, the best option appears to be growing hay, which takes more water than cotton but has more dependable buyers. Miller sells much of his overseas. This is among the most controversial issues in the region, but he calls it a necessity.

We pull over and climb down beside a field of neatly cut, green Bermuda grass. Water washes by in a ditch that's fed through a gushing pipe at the end of the lot. This grass, and, by virtue of photosynthesis, this water, will end up feeding cows and camels in the Middle East. As the ranches and dairies around Eloy pack it in, the local market for hay shrinks. Jace's remaining customers want high-grade feed for show animals and hobby farm livestock, and they'll pay $230 a ton for it. But if the harvest dries out on a windy night or gets soggy from a freak rain, it'll be damaged and Miller will be lucky to make $115. Buyers from China and Saudi Arabia—where growing alfalfa to feed livestock was banned in 2018 to save water—will pay more for the cheap stuff, and they deal in set prices, not the seesawing market value of market whims.

"Don't subsidies help make up the difference?" I ask. "I didn't get into farming to cash government checks," Miller replies. "And disaster relief doesn't make a living."

And yet America's crop insurance program, which has replaced the practice of paying direct crop subsidies in recent years, carries the largest premium value in the world.[49] Pinal's agriculture receives more government assistance than any other county in Arizona. From 1995 to 2020, Pinal farmers raked in $643 million in subsidy and insurance payments, three quarters of which was paid to cotton farmers who blamed rising temperatures for half their losses.[50] Crop subsidies and insurance are a common soft adaptations to climate change around the world. However, research shows that these payments further disincentivize farmers from adapting by switching to drought-resistant crops.[51]

Don't get me wrong, farmers here *are* adapting. The young ones, especially, are reaching for solutions. After all, it'll be they who have to find new work. In the state's oldest citrus orchard, I walked through the bright scent of fresh orange and tangerine and learned how remote moisture sensing and drone monitoring can help farmers irrigate with pinpoint accuracy. I've spoken with tribal farmers who promote the cultivation of beans, squash, and corn that evolved on these lands. And I've peered across acres of an experimental shrub that scientists at the Bridgestone Corporation hope will produce a long-sought domestic source of high-grade rubber.[52] This adaptation is particularly encouraging.

The plant, called guayule, is native to Mexico. Bridgestone's biologists are breeding it to develop a strain that produces latex for use in surgical gloves and valuable resin with far less water than is needed to grow cotton or hay. They've even designed the specialized equipment needed to harvest it. But their next task, which has stymied rubber-growing efforts in America for decades, is to create a market so lucrative that risk-averse farmers are willing to trade their hay and cotton seeds for tiny grains of guayule. This would be difficult even if the region weren't locked into the CAP framework, and it would have the ripple effect of starving local dairies.

~

One afternoon, I was driving through Mesa, heading west along the city's new tech business corridor. A slate-gray wall rose out of the dirt like some alien monolith with construction crews climbing on scaffolding around it, but I couldn't tell what they were building. I turned onto a side road and caught a glimpse of the faded sign for the Rijlaarsdam Dairy. There was room for several hundred cattle in the paddocks but only three sullen-looking cows milling about. A sign over the door said "Tours Daily," so I pulled in, hoping they knew something about the project across the street.

Rijlaars, it turns out, are riding boots, in the Dutch tongue of Glenda Stechnij's ancestors, who came to farm here four generations ago. They hailed from the Netherlands, where dams are so fundamental to life they became synonymous with "city." Stechnij said she now makes most of her money on a small petting zoo that hosts any kid's dream birthday party. Out front, chickens pecked around the entrance to the clapboard shop. Inside, refrigerators chilled frozen custard and cheese, and a small gang of inquisitive kids requested carrots to feed the baby goats.

"We were way out in the desert in 1976. We had to put in our own road, a well, all the infrastructure," Stechnij told me. "The dairies clustered together to make buying feed and selling milk easier. Between them we used to have about twenty-thousand head of cattle. There were eight family dairy farms within a two-mile radius for half a century, and now they're all gone." She was forced to sell her own cows at the height of the COVID-19 pandemic when schools closed and milk prices plummeted and the Dairy Farmers of America dumped 3.7 million gallons of it down the drain each day.[53] She hoped to replace them, but the 1-million-square-foot Amazon fulfillment center going up outside her windows foretold of other troubles.

About five years ago, a coalition of builders began lobbying the city to change the zoning of the area to accommodate new tech firms and a master-planned development. In April 2020, work began on 1,100 acres surrounding her farm. Her tiny 10-acre plot with the petting zoo was exempted from the zoning change. She had another 18 acres to the north but said it was in default awaiting a judge's ruling. And she was being sued for holding out on her portion of a 20-acre plot she co-owned to the south. When her husband died, their business partners began pressuring her to sell while developers were buying by the square foot, but she refused. Now she's embroiled in a legal battle that, to her, is about a lot more than the money.

"It's tempting," she admitted of moving on, "but I love where we live. We built our homes and raised our kids here. I pretend sometimes that the old farmhouse is surrounded by skyscrapers, but I held my ground." The place where we talk stands alone beneath a single tall tree and is surrounded on three sides by weedy desert lots. Housing developments inch down from the north, and Amazon's featureless gray structure lurks on the eastern horizon.

She told me that a Phoenix-based law firm had been pushing to develop these lots, and her story suddenly rang a bell. As I watched the gaggle of kids head out the back door with hands full of goat treats, I recalled Cole Cannon, who was building a beach for his six children on what had been a farm just 3 miles away. He purchased the 40 acres, he told me, while helping mediate a dispute between its previous owners, who had been legacy agriculture people. Over the years, the old family farm had been broken up between the new generations, and Cannon, a lawyer, got involved because of a stubborn holdout. There was "one black sheep who was making life difficult," he said. So, he told me, he threw money at them to make it go away.

For generations, communities like Eloy and Gilbert revolved around the planting season. Today, they confront a fundamental change and a question of societal values. Cutting back what they plant might seem reasonable from the outside, but from within it looks like someone chipping away at your identity. For many farmers, water is a savings account and a retirement fund. As long as they have their right, they are relatively safe. Reduce that, and you threaten more than a season's income.

Across the West, cities are buying cropland or paying farmers to leave fields unplanted. As long as three quarters of the region's water is tied up in agriculture, adapting to drought will necessitate fallowing fields. Farms use as much as eight times more water per acre than three homes.[54] "Buy and dry," they call it, and although it promises to gut rural communities by eviscerating their economies and erasing their culture, it's probably the future.

For cities, there are other benefits besides water. "The transition of a given acreage of land, from agricultural tax base to a residential community, factory or sky rise building, results in the tax rate going up and the assessed valuation going up," said Rick Gibson, who recently retired after forty years as an agriculture extension agent and master gardener in Pinal County. "As the assessed valuation goes up, there's more ability to draw money to build more infrastructure to fuel growth. So, because agriculture uses quite a bit of water, there is an expectation that agriculture will go away and everybody else will have access to water resources and continue to grow."

Older farmers confront this reality toward the end of their career, but Miller is just getting started. "Most of my good buddies, we're kind of digging in our teeth and just trying to hang on, hoping that we'll get a wet, wet winter. Always praying for rain," he says. "Even if that happens, I just have this ungodly gloomy feeling."

In an ironic twist of fate, one potential result of Anglo farmers losing their river access is the tribes reclaiming their previous role as the region's agricultural powerhouse. They have the water, but their farms sat idle for so long that few tribal members have the capital or experience needed to operate them, and infrastructure remains sorely lacking across the reservation. As a result, only a fraction of its fields is ready to be planted. This has presented Miller with a unique opportunity. He has what the tribes lack, and so they have partnered. His business helps plant and harvest hay on half the 10,000 acres that the tribal nation owns, plus more acreage for individual growers. It's a good arrangement, but there's no telling how long it will last. The Gila community has more water than it can use now and leases some to cities like Phoenix, but big cuts in river deliveries are coming, and no one is immune.

"At the end of the day we all want to work together to solve problems," said Stephanie Sauceda-Manuel, who manages the tribally owned

Gila River Farms. "I work with farmers outside the community and I see them struggling, selling their land and losing cattle. There are houses going up all around us, and that's taking away from the small farmers. I think that's sad."

Miller harvests all of her alfalfa and Sudan grass. "Jace has been true to his word and his work and his commitment to the farm, that's why he's been around so long," she told me. She plans to cut back on cotton acres and focus more on her top seller: alfalfa. "We hit ranching, farming, and rodeos are big business for us. Some of our stuff goes over to China and Japan. That market is going to be huge for us in the future, especially because we're going to be losing the smaller local ranchers who we sell to now."

Miller knows what's coming, too, but he plans to keep driving, keep farming, keep trying. "The only way we stop is if they're putting us six feet in the dirt, or we declare bankruptcy. I don't know if we're that resilient or that ignorant."

I leave him with Lexi and another farmer inside his sparse brick house and walk out into the night. They have a mini-fridge full of Bud Lights and a long, hot day of mishaps to discuss. Out in the gravel driveway, the sun has fallen below Picacho Peak and three diesel trucks idle in an unattended purr. For a moment, I think about going back inside to tell them their cars are still on. Then I find the sense to drive away.

I roll past empty lots with signs seeking new investors, skirt the active adult community with golf greens that shine iridescent in the night, and pull onto Interstate 10 northbound. The highway tightens with traffic as the glow of Phoenix rises. Even late, there's no relief to the churn. Work crews in neon orange vests tear into concrete under blinding white lights. Billboards announce new communities, fun times. Tricked out Subarus split lanes and heavy-duty Dodge Rams tailgate at 90 miles an hour. Critics have been foretelling Phoenix's demise for decades—willing it, even—and yet the place just keeps on chugging like some immovable force, like a diesel truck left idling.

Eloy is well behind me, out of sight and out of mind, but what's happening there has implications for us all. The collapse of central Arizona's farming communities seems nearly inevitable. By using farmers as a means of securing water for future urban development, the CAP effectively ordained it. Today, prisons are Eloy's new job provider. The town has four of them, which ranks it thirteenth nationally per capita.[55] Quite the change from being among the country's top cotton producers. The town emerged from the desert soil and grew piece by teetering piece upon a precarious system of diversions, dams, reservoirs, laws, contracts, and canals—always expanding on the promise of the resources it would make available. Now, it's come rushing headlong into the realization that this assumption was dead wrong.

I can't help wondering: What makes us think things will turn out any differently for those in the glimmering city ahead who are embarking on the latest quest for El Dorado?

Long ago, I asked the director of a climate research center in Tucson whether the state's ambitions were pushing the desert beyond its carrying capacity. He answered that the term no longer applies here, because we don't depend on our environment for survival anymore. I don't know whether we're that resilient or that ignorant, but I'm sure that building a surf park in a desert where worsening drought is condemning future generations to a future defined by water scarcity is, if nothing else, a very bad look.

The Stories Caves Tell

Far out in the wastes of Nevada, hundreds of miles from any place most people would bother to wander, there's a yawning hole in a dark hillside at the southern end of a narrow mountain range. The hole appeared suddenly on some forgotten day when the hillside gave way, exposing a chasm 75 feet across. If you look up from inside, sawdust colored limestone walls rise over a sloping floor and curve inward toward an opening like an eye at the center of a crowned ceiling. Beneath this opening, green bushes cluster in the shifting footprint of the sun's penetrating light.

In the early hours of one summer morning, a rope drops out of the hole's opening and unspools 15 feet into the chasm below. A figure leans out over the edge against the cerulean blue sky and rappels down the rope until his boots hit bottom with an echo. Matt Lachniet, forty-one years old at the time, brown-haired and bearded, bends backward to take in the entrance of Leviathan Cave. Lachniet wears a red helmet and carries a small pack. He unhooks himself from the rope and walks toward the outer edge of the sinkhole, crossing into the darkness beyond the thin diaphragm of sunlight. When he ignites an electric headlamp strapped to his helmet, opalescent feldspar, pink quartz, and obsidian Apache tears glint to life in the walls. He finds an opening about the size of a small Hobbit hole and stoops to his knees to crawl inside. Behind him, the sun has reduced to a horizontal slit; before him is nothingness. He follows the rough tunnel deeper and deeper and then is gone.

The air inside the cave's inner cavern is absolutely still and stale, like a Cold War–era bunker. The room is small, and its ceiling drips down in long strands. These stalactites glisten in his light as they reach for stalagmites below. The whole of the cave is shelled in what looks like dried wax polished to a gold luster. Lachniet searches for a "dead and down" speleothem, a stalagmite that has toppled over onto the cave floor. When he finds what he's after, he places it into his pack and turns back toward the distant light.

In 1929, a Vermont astronomer working in Arizona made a scientific breakthrough that changed how we understood Earth's past. Andrew Ellicott Douglass was trying to track the effects of solar variations on

Earth's climate when he recognized that trees added layers of growth with each passing year and that those layers, which can be dated, are influenced by environmental conditions. In wet years, tree rings may be thicker than in dry years, for instance. He applied his tree-ring dating method—dendrochronology—to prehistoric cliff dwellings using wood found in abandoned ruins and determined the dwellings to be roughly 2,000 years old. As it happened, their abandonment coincided with a great drought in the late 1200s ce.

The science has since advanced so that researchers at the Laboratory of Tree-Ring Research at the University of Arizona can use cores pulled from ancient juniper and pine to look 12,000 years into the Southwest's climatic history. And policymakers planning for the current drought have always relied on this information. However, in the grand scheme of things, even 12,000 years is a limited window.

During her tenure as Phoenix's director of water services, Kathryn Sorensen helped to steer her city away from imminent catastrophe. But that didn't mean it was out of the woods. "We are prepared for shortages, but we are not prepared for a worst-case scenario," she told reporters in 2017.[56] The worst case managers often point to is a ten-year period between 1146 and 1155 ce that occurred during a drought-ridden stretch of the Middle Ages. Water managers have used this benchmark to design infrastructure, bank groundwater, and secure surface water for the future. But it turns out that this period is only a sample of the potential pain. For a real worst case, we have to look back further. Luckily, caves tell a longer story.

Lachniet is a paleoclimatologist with a government-issued permit to collect cave samples solely for scientific research. He uses samples taken at caves across the American Southwest to understand how the environment has changed through time. Back at his lab at the University of Nevada in Las Vegas, he slices the speleothem into thin strips like pages of an open book and studies them for clues about ancient droughts.

His findings are revelatory. Farmers have told me they are prepared to weather a fifteen-year drought. Urban water managers say they've planned for decades. All these estimates pale in comparison to what Lachniet has learned in Leviathan Cave.

All alone in an arid swath of the Great Basin, Leviathan has remained mostly stable through the ages. Outside, meanwhile, the region's climate has fluctuated between dry and wet periods in a relatively consistent cycle lasting about 21,000 years.[57] Inside the cave, these changes have been recorded as water seeping through fissures and adding layer upon layer to the cave's stalagmites, much like tree rings. By carbon dating speleothem growth, Lachniet parses out trends in monsoon and winter rains dating back 175,000 years. His data show that the region began warming around 11,700 years before the year 2,000 (B2k) and that warming peaked at 8,380 B2k. What matters most is the length of time—the 2,180 years between 9,850 and 7,670 B2k that saw especially slow stalagmite growth rates, reflecting an especially warm and dry period.

In short, cave records help prove that the potential worst case facing the West is not a decade-long megadrought as seen in the twelfth century but an epoch of aridity lasting ten times longer than our country has existed. While Lachniet explained this to me, I remembered Yuichi Ebina's weathered family records and his proof that giant tsunamis didn't wait 1,000 years to batter Tohoku. How do you plan for what you don't know is coming?

"Visually, the Middle Holocene might have looked pretty similar to today," Lachniet told me. "It's even possible that the summer monsoon was a little stronger then than it is today." But summer rainfall does little to replenish aquifers, because so much of it either evaporates or is sucked up by plants. Winter is the season that matters. That's when temperatures are low enough that rainfall seeps into the ground before being lost to the sky and, crucially, when snow sticks to the mountains.

The Proto-Uto-Aztecan people who lived in the Sonoran Desert at the time hunted and gathered successfully enough to establish a consistent presence 2,000 years before agriculture took root here.[58] During this long stretch of dry years, Lachniet's records show they would have had very little winter rain to speak of, and many archaeologists suspect they mostly abandoned the place. Tracing the limbs of language trees reveals that some went northwest into the Great Basin and others retreated south to the foothills of the Sierra Madre Occidental.

Twenty-five-hundred years of heat, dry winters, sparse game, and parched streambeds seemed bad enough to me, but what rattled Lachniet is the likely cause. The motivation for this ebb and flow is celestial. Insolation, or solar radiation received at Earth's surface, changes with the Earth's tilt and position around the sun. The warming that occurred during the Middle Holocene could not have been a direct response to solar radiation, however, because it lagged behind what our records show of insolation by a couple thousand years. There are a few machinations of the Earth's climate system that leave a similar fingerprint of warming on the heels of increased solar radiation. One of those is a change in the Arctic. More sun at the pole can cause a gradual loss of Arctic sea ice and a warming of Pacific waters. Over a couple millennia, sea ice loss can disrupt atmospheric circulation within and over the Pacific Ocean, which contributes to the weather patterns that bring moisture to the Southwest.

So why does Lachniet find this so troubling?

"If a link to the Arctic operates in the future as it did in the past, we are rapidly on our way to return to Middle Holocene conditions," he explained. Over the past couple hundred years, our species has been doing to the Arctic what the Earth's tilt takes several thousand years to do. We have watched the ice sheet calve in Greenland, glaciers recede across Alaska, and sea ice at the pole recede dramatically. These changes might seem far afield of the Saguaros guarding the Sonoran Desert, but they are not.

"Over the last 2,000 years, there were not huge variations in global climate. It's been relatively stable. That's important because water policy people have only been focused on past decades or even millennia. The longer we go back in time, the bigger the variations we see," Lachniet explained. "For the last sixty years, we've been relying on tree rings to understand the Southwest; they only go back a few thousand years. What we find when we look at these other archives is that we're missing a lot of what nature is capable of doing. The whole story is more extreme." It's a story that casts doubt on assurances that desert communities are prepared for what's coming, or even that they *can* prepare for what's coming.

Simply put, "We're going to have a reduced supply of water. There's not enough water for multiple Las Vegases and all the agriculture, but right now it's being met with temporary and minor reductions," he said.

I wonder how long it took those prehistoric people to move on. How many generations ventured farther from familiar lands in search of a reliable stream before they left altogether? How long did they squat in the dirt, peer up at the empty sky, and wonder what happened to the rain? They couldn't have missed the changes occurring around them, so tuned to the landscape as those who forage and follow game must be. And this is what *I* find so troubling about what Lachniet learned in Leviathan Cave.

As he goes on about speleothem and isotopes, I feel a churning deep down in my guts. It's not just the likelihood that we are entering a millennia-long period of drought that turns my stomach, though; it's realizing that our modern amenities have made us so blind to it. We have more scientific data about our home than ever before, but we don't live off the land anymore. Most of us hardly interact with it and wouldn't notice if a wave was bearing down on us from out there in that empty desert. That's partly because, much like seawalls, dams and canals have blocked our view.

A Thousand Cuts

I slide my driver's license through a small hole in a thick glass window so the attendant behind it can record my data. Then I enter a man trap, locked between two heavy metal doors with a camera monitoring my nervous shifts. I'm stuck until Bruce Barnes leans forward to scan his keycard on an electronic pad. It beeps cheerfully, and the door clicks open. We enter a long hallway with clean, white walls and a glassy floor illuminated by frosty blue track lighting. Paper chandeliers hang in puffy clumps like clouds from the exposed ducting of the industrial-chic ceiling. The twang of Stevie Ray Vaughan's guitar tinkles from some unseen speaker as Barnes leads me past suites of computer servers that house untold quantities of the world's data.

There are at least thirty-six data centers in the Phoenix area.[59] This is H5, located in Chandler, 15 miles by car from Cannon Beach. Covering 180,000 square feet and using 26 megawatts of power, it's certainly not the largest but may be among the most advanced digital storage facilities in a metropolis profiting mightily off the spoils of the latest tech boom.

"Chandler was so welcoming," Barnes says. He's an electrical engineer with icy blue eyes and closely trimmed dark hair, and he manages this digital repository. "This was an alfalfa field, but all the carriers were already in the street, so it was trivial for me to bring in additional bandwidth. Without the bandwidth, you don't have a data center." I recall being told that back in the 1970s Chandler had gifted a parcel of land to Intel so it would build the area's first computer chip manufacturing plant. "That doesn't surprise me," he says. "Chandler was cow country then, so the land was plentiful. When we built this, there was a dairy farm right outside. There were farms all over. Their owners are all millionaires now."

In 1862, President Abraham Lincoln signed the Homestead Act, which offered 160 acres of territorial land to anyone who could pay a nominal fee, move West, and start farming. It cracked the whip for entrepreneurial families like the Millers and kicked off a land rush. Today, cities like Chandler and Mesa lure tech moguls with ample land, abundant fiberoptic cable, and lucrative tax incentives. Arizona's data centers pay no state, county, or local sales taxes on equipment purchases.[60] Tech companies also choose the Phoenix valley for its comparative rarity of hazards such as earthquakes, cyclones, and tsunamis.

Intel recently broke ground on two new fabrication facilities at a cost of $20 billion, the largest private sector investment in state history.[61] Others have followed: Motorola, Taiwan Semiconductor, Microchip, ON Semiconductor, VLSI Technology, Freescale Semiconductor, NXP, STMicroelectronics, Honeywell, Retronix Semiconductor, Maxim Integrated Products, Marvel, Amkor, Advanced Semiconductor, Analog Devices, Jabil Circuit, Infineon Technologies, FlipChip, Philips, Western Digital, Medtronic, Dialog Semiconductor, Lansdale Semiconductor, and so on. Because Chandler is about tapped out, Barnes says new investment has shifted to east Mesa: Apple's 1.3-million-square-foot global command center. Google's planned 750,000-square-foot data center. A 960,000-square-foot facility whose mysterious owner kept locals in a tizzy before unveiling it as Facebook. Huge warehouses for tiny bytes.

This explosion of investment would not have been possible without the water and stability that the CAP's arrival promised. As these facilities and the homes that follow closely on their heels extend across the landscape, they give the Phoenix megalopolis its characteristic amorphous shape. The scale is best appreciated from a plane or time lapse of satellite imagery that shows the browns of desert and greens of farmland giving way to the gray monotony of a digital mecca. Unlike those earthy tones, however, once the city has grown—and increased its water use—it cannot pull back so easily.[62]

Planners have a brief window to avoid committing generations to wrenching decisions like those Miller now faces. So far, though, the state has not produced a cohesive adaptation strategy to guide those choices.[63] From the tea leaves in tree rings and speleothems, it appears that urban populations are sleepwalking into the same mistakes as the farmers: downplaying warnings, prioritizing growth over all else, and believing that they can build themselves out of any hole.

"I have deep concerns about our long-term needs for Arizona in general," Mesa's vice-mayor, Jenn Duff, told me when I spoke with her before visiting H5. "Mesa has a healthy and diverse portfolio of water, I don't think we're in any immediate danger, but our decisions affect the whole."

Tall, fit, and blonde, Duff gave off the air of someone who takes no shit, and she has shown little fear of voicing concerns others on the City Council mulishly avoid. Born in Mesa, she toured the world as a professional bass angler before starting a fashion business and then buying a bungalow in the city in 2009. Mesa was terribly depressed back then, but she saw potential. "I fell in love with my little downtown and wanted to bring it back to the days when I was little girl and it was the place to be," she told me.

Today, downtown Mesa holds tentatively to that old-timey charm while juggling new investment projects dominated by technology firms and Arizona State University. In her municipal role, Duff lobbies for infill development that builds in vacant lots and repurposes old buildings rather than pushing farther out into the virgin desert, but there's little stopping the sprawl. "We've seen phenomenal growth on the east side. The city built a tech corridor, making sure we had the infrastructure needed to attract those tech companies," she said. Now she harbors serious reservations about the prudence of that effort.

Tech manufacturing plants and data centers, in particular, demand prodigious amounts of water, and the industry depends disproportionately on highly stressed watersheds such as the Phoenix valley.[64] Their processors and servers whir ceaselessly through day and night, producing one thing the desert has in abundance: heat. To keep them from frying, many facilities push up to 1.25 million gallons of water through liquid cooling systems each day. "This is an irresponsible use of our water," Duff complained. Mesa could not become semiconductor central without the CAP canal, which provides more than half the city's water at a cost of about $8.7 million each year. As the Colorado River dwindles, that price is expected to balloon 40 percent by 2026.[65]

Many of the data centers I contacted were reluctant or unwilling to let in a journalist. The drought has brought scrutiny. Barnes agreed to show me around, perhaps because his facility is something of an outlier. I came to Chandler to witness the gluttony of the latest tech surge, and I did, but as he opens one of the doors to expose a suite of servers and invites me inside the narrow space, I'm confronted with a rare pang of optimism. "Everything here is dry," he says. "We aren't using any water to cool. The only water we have is for humidifiers, because we don't want electrostatic discharge." Is that common? "Others use big evaporative coolers, and they consume lots and lots of water. There's some advantages to that, but there's disadvantages."

The sweet spot for hardware is 82°F, and as Barnes speaks our body heat raises the temperature in the small space enough to trip the cooling system. Overhead, a fan whirrs to life like a helicopter lifting off. The room gets cold fast. If you're not using water to cool, what are you using? Electricity, of course. Collectively, the nation's data centers consume as much energy as New Jersey.[66]

~

As the drought drags on, the region remains mired in debates monopolized by the Colorado River, as if water were the only issue. It is not.

Desert city streets that once echoed with the clomping of hooves now buzz with the sixty-cycle hum of air conditioners, damn near an

A-flat that thrums on like a one-armed guitarist with five broken strings. Running these machines all day and night has helped cause peak energy demand to double between 2009 and 2014. Three quarters of that power was generated by burning natural gas and coal.[67] Nuclear production came in a distant third behind hydro, the future of which grows dim as Lake Mead falters. Despite the desert's sun exposure and claims that energy here is practically free, all renewables account for just 10 percent, thanks in large part to utility companies—which have little to gain from residents generating their own power—that have consistently outspent their opponents to quash renewable energy initiatives on the ballot.[68] And so, as more people move to the desert, more dirty fuel will be needed to make their lives comfortable.

In the summer of 2021, even that energy faltered. During a fresh wave of COVID-19 infections, with more people working, streaming, exercising, and mining Bitcoin at home, Mesa nearly ran out of energy. "We have an economic explosion and more demand for electricity. It's been hard to find enough electricity," Duff told me. In response, Arizona's third-largest city asked its residents to set their thermostats 5°F higher. "The energy prices are astronomical, like nothing we've ever seen before, so we asked people to conserve as much as possible," she said. The loss of the Navajo Generating Station's cheap coal hasn't helped, and how Mesa will keep power within reach of low-income residents in the future is unclear. "It's truly been a challenge. If we do aggressive price increases to keep up with what we're paying for electricity, we're going to have more problems with heat-related illnesses and deaths," Duff said.

From 2008 to 2018, heat-related deaths in Arizona rose from 112 to 251 per year. Victims were disproportionately Native American, Latino, Black, and elderly, and one third of them lived on the streets.[69] City health officers have responded by designating air-conditioned public spaces like libraries as cooling centers, but these are hardly up to the task, especially when multiple hazards collide, as they did in the summer of 2020. That year, Maricopa County closed 87 of its 106 cooling centers because of the coronavirus outbreak, and extreme heat killed at least 510 people in Arizona, including 315 in Maricopa.

Devoid of form, taste, sound, or smell, heat secretly infiltrates every corner of our lives. It slows our cognitive function, causes insomnia, and triggers psychiatric illness.[70] As temperatures rise, so do reports of violence and mental breakdowns, especially in urban heat islands devoid of green space. Suicides tick upward with the thermometer. High temperatures decrease the efficiency of electricity generation, coal-to-gas conversion, turbines, and solar photovoltaics. And it lowers the energy capacity of power lines and transformers, threatening the function of air conditioners in sweltering homes.[71]

In 2022, Maricopa residents rode out 145 days with temperatures above 100°F.[72] There's no reason to expect that number to go anywhere but up. In Mesa, the city council has discussed responding with what Duff called "creative things" that largely amount to planting more trees and opening more public cooling locations. Meanwhile, it continues to roll out the red carpet for water- and energy-gulping data centers and the workers who follow.

"As a state we have a lot of economic development measures and programs. It's time as a state, and as a city also, that we start marrying our economic development goals with our natural resources," Duff told me. "We can't just go on with economic development and then question whether we can support it after they're here." But that appears to be what's happening. While water gets all the attention, a changing climate also brings scorching temperatures, erratic rainfall, gusting winds, landslides, and—all of which pose immediate threats to residents.

In the low-lying, dusty Phoenix valley, air quality is a particular concern. The Arizona Department of Environmental Quality publicly acknowledges that concentrations of airborne particulate matter in Phoenix are "occasionally above federal standards." In fact, Maricopa County has not attained US Environmental Protection Agency (EPA) air quality standards for twenty-three years.[73] Now it appears to be getting worse.

Katie Clifford, a geographer at the Western Water Assessment, has uncovered how health officials quietly obscure this fact through a loophole in the Clean Air Act that allows the exclusion of exceptionally bad air quality events from the data they submit to the EPA. Using the rule, Maricopa has reported no more than three exceptional events a year (the EPA's allowance) since 2007. In truth, some years saw twenty-two exceptional events, but nineteen were omitted as outliers. By classifying as exceptional measurements from days when, say, a haboob engulfed the valley in a swirling mass of paint-stripping grit or dust from fallowed farms mixed with highway smog, the county may avoid stricter regulations and fines.

Although entirely legal, this omission blurs the line between a good day and a bad day. As we go stumbling into this no-analog future, a question emerges: What *is* exceptional?

My mind darts to the hollowed-out towns of Fukushima and the engineers who excluded the worst tsunamis from their coastal protection plans; the East India Company men who ignored the link between siltation with embankments; and Eugene Clyde LaRue, who showed Colorado River flows to be exaggerated. Shuffling bad air quality data away as an outlier creates a false sense of rarity; it avoids acknowledging the true risk and the need for more complex adaptation. In effect, it leaves millions unaware that they live in the path of a slow-moving disaster.

The threat that the desert faces can't be seen creeping up a sandy beach or throwing sparks into the night sky as it devours pines. Drought envelops communities gradually, bleeding them dry over generations rather than washing them away in an instant. It undercuts agricultural earnings and increases the cost of infrastructure, housing, and transportation. It causes rolling brownouts that leave the infirm to die in overheated apartments and mires planes on melted runways like flies on sticky paper. Lettuce pickers fall dead in fields, and people go mad in the streets. Far from a discrete, Earth-shattering catastrophe, drought is a death by a thousand cuts.

The dams and canals that bring water to the middle of the desert have not directly caused these hazards, but they certainly do nothing to ameliorate them. However, they do lure more people to where their effects will be felt.

Domus Operandi

One afternoon, I drive east out of Gilbert and keep going so far that I pull over to make sure I haven't missed a turn. Then I drive some more. I pass endless concrete block walls that conceal all but the pitched tile roofs of hidden tract homes and green strips of well-watered grass whose only function is delineating one lane of asphalt from another. I circle pampered lawns that welcome me to planned neighborhoods with gushing fountains and ponds so artificially blue I wonder if the dye has created endemic flocks of periwinkle mergansers. In east Mesa, far out on the fringe of the Phoenix megalopolis, the only vertical relief is the Superstition Mountains. Otherwise, it's all houses, and the houses are about the same height, like stalks of corn.

Growth has always been the goal. The vehicle may shift from agriculture to tourism and tech, but the aim has been selling houses and bloating the tax base to fund more highways, stadiums, and continued growth. Farmers cry that if they're pushed out the state will lose a $23-billion agriculture industry. Construction alone brings $15 billion to Arizona.[74] Real estate, rentals, and leasing bring another $50.6 billion. In 2020, before the COVID-19 pandemic put the kibosh on housing development across the country, the number of annually issued building permits in Arizona rose by 26.49 percent, blowing the national rate of 6.59 percent out of the water.

An acre of cotton brings a dependable annual return. An acre of houses filled with consumers promises lifetimes of compounding value. Buyers just have to be convinced that all is well. In that, the CAP has been key. Even as media outlets from Tucson to Dublin report a looming catastrophe on the Colorado River, people continue to move to places that depend on it. There's a clear disconnect between the apparent danger and the associated concern, and it can be traced to the 165 miles of desert separating the river from Phoenix's newest surf park. Unable to see the source of the trouble, and conditioned from years of standing aside while experts debate arcane water law, we've come to accept a distant risk that, we're told repeatedly, poses no immediate threat. A drive through Eloy says otherwise, but who would bother?

I think of Konno-san's wife, who died behind her seawall, and his shame for having not been there to save her.

Far out in east Mesa, I enter a master planned development spread across some 3,200 acres with clumps of model homes and rows of empty dirt lots. This is Eastmark. Once finished, it will include 15,000 houses, of which only 4,000 stand when I arrive. From a bird's-eye view, this is the largest and most uniform housing development project in the area. It oozes like spilled gray ink between the Mesa airport to the west and the open desert to the east. It's the work of a coalition of builders with their own flavors, but the houses all look about the same, and nothing in the layout makes any sense to me.

The street names might have been chosen at random from the glossary of a picture book about outer space: Sonic, Sunspot, Isotope, Carbon, Stellar, Flare, Rubidium, Retrograde. I idle at the corner of South Copernicus and Solina Avenue, in the middle of an empty street surrounded by empty lots and empty houses. There's something unsettling about the place, like I've stumbled into a testing site for the Manhattan Project. There are no prisons nearby to employ the thousands of families filling these homes, and I wonder what's bringing them all here. I need not travel far to find out.

Eastmark Parkway, the community's four-lane, sclerotic spine, ends abruptly at a chain-link fence topped with glistening barbed wire and watchful security cameras. I peer through the metal links across a vast construction site where a few backhoes work over 300 acres of brown dirt like toy trucks on Long Beach. Facebook is spending $800 million to transform this dirt patch into a massive data center, the latest addition to Mesa's Tech Corridor that already includes Apple and Microsoft. At Eastmark, children will play in the shadows cast by their parents' employers, not unlike the ghosts of those that haunt derelict clapboard shacks in busted mining company towns all across the West.

The luxuriousness of these homes, the greenness of their outdoor spaces, the smiles on the billboards, it all engenders a sense that this will last. On Solaris Street, work crews nail up another particleboard house against an azure sky. On Sector Drive, a realtor in a flowery dress strides out of a manicured model home with a frumpy middle-aged couple in tow. On South Ozone, a man and his daughter pedal their bicycles past the image of a smiling multiracial family that beams from some builder's advertisements. I approach the nearest model home, which appears representative of the whole. It's a broad ranch with gabled tile roof that boasts 2,833 square feet, four beds, three baths, a study, an "owner's retreat," and a four-car garage. The exterior walls are light gray stucco. The roof is dark gray colonial tile. Faux gray rock wraps the base. For no apparent reason, they call it the Roadrunner, and they sell it for $799,990, which is about average here.

I walk inside and am immediately dwarfed by the executive ceilings. I'm enamored with the slab marble kitchen island and have to pick up my jaw in the enormous dressing rooms. The place is gorgeous, no doubt. What really gets me are the windows. In the central great room alone there are seventeen, all facing south or west, as well as three sets of sliding-glass double doors. In midafternoon, the Roadrunner is absolutely steeped in natural light. This is a unique feature in the many-thousand-year history of desert dwelling.

To be in tune with the desert is to be on the lookout for shade. A road-runner sprints from the cool of an overhanging rock to the dim reprieve cast by a jojoba. As a pale kid growing up in the desert, I learned to do the same. The afterschool ride home in the un-air-conditioned bus was a stifling misery of sweat and smog that ended beneath a slope of baking asphalt. Head down, I trudged to my house, broke through my front door, peeled my soaking shirt off my back, and collapsed on the tile floor, thankful, at last, to be hidden from the sun. Summer was the slow season when we waited out the searing afternoons in caves of cool darkness watching TV and harassing our sisters.

The first wave of housing designers understood this. They built squat ranches from clay or painted slump brick and topped them with flat roofs that stretched over a few inset windows intended to stay dark. In this, they picked up on what people indigenous to the desert, not to mention the colonial Spanish, had learned through practice. Structures that predated modern air conditioning relied on shade to beat the heat. The classic colonial casita is encircled by a thick adobe wall that butts up to the road and rises overhead as a first line of defense against the sun. Shaded and insulated, the world within is an open-air oasis with living quarters and often a rain-fed garden.

European Americans' adaptations to Arizona have long sought to meet Eastern expectations. Tucson had just three trees in 1875.[75] A year later, it received its first bunch of Bermuda grass from San Diego. By 1910, following a concerted effort to make the inhospitable landscape inviting, thousands of exotic species spread over its horse-trodden dirt roads. Tucsonans planted so many trees that they sucked the ground dry, killing off much of the native riparian woodlands and mesquite *bosques* along the Santa Cruz River. That's when they first turned to groundwater, plumbing a 50-foot well to irrigate a stand of cottonwoods planted to welcome easterners off the Southern Pacific Railroad. Ever since, we settlers have put more effort into molding the desert to our liking than the other way around.

The buildout of centralized utility systems that tapped into energy once deemed "too cheap to measure" and "renewable" water delivered from an unseen source have erased all need for developers to build with the desert in mind. So they don't. With air conditioning, modern homes mimic New England's shingles. They are multistoried, carpeted, and capped with gabled roofs waiting to shed snow. They're packed together on tight lots surrounded by heat-trapping asphalt and concrete. And they're huge: Paradoxically, while the size of the average American family has fallen, homes have swelled. The average house built in postwar Tucson was 1,200 square feet; the basic Roadrunner in Eastmark is nearly 3,000 square feet, and it's hardly the biggest one on the block. Every extra foot is more space to heat and more to cool. More to light and more to clean.

Then, of course, there's the lawn. Though modern developments maintain some of the classic casita's adobe styling, they sit back from the street to make room for this new feature. Much has been said about lawns in the desert, which have always looked to me like a poor man's imitation of an English bowling green, and cities around the Southwest are finally standing up to this antiquated, alien design flaw. Las Vegas, especially, has led the way toward encouraging or forcing residents to tear out their turf, and the trend has spread, but there's still a long way to go. Even in comparably environmentally minded Tucson, city ordinances that incentivize using graywater or rainwater for landscaping have failed, probably because of poor communication and lax enforcement. As a result, 30 percent of the city's drinking water goes directly to landscaping.

Inside the Roadrunner, a strikingly tall, coiffed blonde realtor highlights the high-performance wall insulation, which, she says, will keep the energy bills down. Developers are keen to new residents' worries about the cost of keeping these extravagantly spacious houses cool, but as with so many modern adaptations, the concern they seek to address is price. Nowhere in the pamphlet of included features and purchasing information is an acknowledgement that the power might not last—that the electricity needed to drive the Kichler premium designer interior lighting package and built-in ceiling sound system is provided by dams behind which enormous pools of water that took decades to fill and are draining so quickly that they may no longer generate power. Nor do they specify the source of the water that trickles through the backyard fountain and fills the luxurious walk-in bath, and in my experience, few selling or buying these homes know what it is. It comes from two places: the Colorado River, as delivered via the CAP canal, and as many as six groundwater wells.[76]

This is the Wild West of water use in Arizona today. After the CAP came online, the state tried to make good on its commitment to protect its remaining water by enacting a rule that requires developers looking to build new houses within heavily populated urban areas to prove they have a supply to last the next hundred years. Since then, this assured water supply policy has backfired. Developers have simply shifted outside these active management areas to where they can skirt the law. Out here, they pump groundwater nearly at will where the water table is most vulnerable to collapse, and the state did next to nothing to rein them in until the spring of 2023.

"Arizona was famous for having one of the more progressive groundwater management acts of the time and for generally being at the forefront of water policy," said Zach Sugg, a water policy expert at the Babbitt Center for Land and Water Policy in Phoenix. "But the whole thing has just stagnated since the early 2000s. There hasn't been any sort of new innovative policy. They haven't solved any of the existing problems. They're just resting on their laurels," he said. "The intention behind the assured water policy was to put some sort of upper bound on outward sprawl, but when the statute was being written, the developers had influential political sway, and they got some favorable loopholes into the law that allowed them to develop in areas where people want to live, rather than areas that have surface water supply." The result has been the continued construction of groundwater-dependent housing at the edge of town.

All this is to say that despite how it may look beachside in east Mesa, Arizona's groundwater situation—what it depends on as the Colorado River dries up, and what the CAP was supposed to protect—is far from under control.

Water managers across the West applaud themselves for having decoupled population growth from water use, meaning that, thanks to conservation efforts like low-flow toilets, graywater reuse on golf courses, and cutting back on some grass, there is no longer a direct and equal correlation between more people and more water use. Consumption *is* rising, though. Since the CAP arrived in Maricopa County in 1985, annual demand there has grown more than 70 percent, from 633,501 acre-feet to 1,081,451 acre-feet in 2017.[77] The people filling new households don't just flush toilets, water lawns, take showers, and do the dishes. They also wash their cars, eat out, grab a drink, run at the park, go to school, attend football games, play golf, and, apparently, surf.

"I'm definitely not somebody who thinks population is the cause of environmental problems. I will always push back against that," Sugg told me. However, "relative decoupling can never be absolute. There will be an inflection point where you've maxed out all of the available options in terms of efficiency and conservation in your system, and if people keep moving in, there's a point where that downward trend in water demand eventually hits that inflection point and starts going up in the aggregate."

Cities have made laudable gains in conservation, but at the end of the day there's only so much they can do, because the whole state is parched and the whole state is hogtied to a much larger system. One of Arizona's greatest achievements is the replenishment of its depleted aquifers. That was accomplished by dumping Colorado River water into the ground and by transitioning the heaviest users—farmers—off groundwater. In this way, the CAP saved groundwater but not through meaningful adaptation. It simply shifted the burden to the river.

Now, as more homes are built on the edge of town, experts say that there won't be enough water available from the CAP to replenish the groundwater needed to supply those slated for suburban Phoenix.

"If population isn't causing this problem, what is?" I asked Sugg.

"It's not the number of people, it's how the people live. It's the way we build our communities," he replied.

It's huge homes with green lawns and seventeen south-facing windows at the outer reach of utility distribution in a heating, aridifying desert that are depending on the loss of their only local food supply to save them from a water crisis that's baked into a century-old interstate legal compact.

We can thank past adaptations for allowing the illusion of abundance that has made this development possible. And we can blame the economic requirement of constant growth for driving new risks. It's been twenty-two years since anyone tried to pump the brakes on Arizona's characteristically inefficient sprawl. Back then, the Sierra Club–led Citizens for Growth Management initiative dared to draw boundaries around desert cities to set aside natural open space and require that developers shoulder the costs of running utilities to new communities beyond the limit.[78] Their idea was something like Colorado's Open Space program, which contains cities within undeveloped natural lands.

But in 2000, at the moment when a long-shot victory appeared within reach, one of the club's high-profile supporters threw in his lot with the development lobby. And that was that.

In 2021, the state's own water wizards challenged the assumption that precious groundwater was being protected. In a scathing report, they found the state's efforts to safeguard aquifers under Phoenix, Eloy, and Tucson insufficient to ensure long-term sustainability. As Colorado River water dwindles, central Arizona will have to rely on groundwater, and although smart-use policies have pushed residents and businesses to use water more efficiently, "the increasing water demands of the municipal and industrial sectors will continue to outpace the limits of conservation," wrote lead author Kathleen Ferris, senior research fellow at Arizona State University's Kyl Center for Water Policy. "The startling amount of groundwater that has been allocated under assured water supply determinations should give every Arizonan pause."

In practice, as the real estate agent and I walk and talk through the halls of the Roadrunner, the issue of water doesn't even come up. And that's pretty common. When I toured another new development near Eloy, an agent explained that buyers were mostly retirees and out-of-towners who came for jobs at the nearby prison. I asked if any of them expressed concern about the area's water.

"I don't get a lot of concern about water," she replied. "Why, is it bad here?"

Welcome to Dystopia

Carson Miller was born in November 2021. "A lot of this has changed since that little boy came into the world," Jace told me of his son. "I want to leave him with something that he can grow into, and I want to leave something better for him. I have generations of knowledge from my family doing this here. I want him to be able to use it, if he wants to." I asked him what the key to staying in business is.

"We have to start small, start simple," Miller answered with characteristic honesty. "We need more fucking water. That's a simple premise. You look at the Mississippi; it causes one to three million dollars a year in damages from flooding. That river just likes to tear shit up," he said. Someone should go bring it to Arizona.

It's been a mere fifty years since the state went all-in on outside water to save its aquifers and find a sustainable supply for its future. With neither of those goals fulfilled, it's gone searching again, because the CAP didn't just set a precedent for distant water grabs, it necessitated them by setting a standard for growth that it could not maintain.

Across the Mississippi River system in 2019, flooding from winter rains, spring storms, and summer hurricanes cost more than $20 billion. Soy and corn farmers took the brunt of it. A world away in Arizona, Miller faced losing his farm for lack of water. In an eerie resemblance to 1980s Bangladesh, the wife of Casa Grande councilmember Dick Powell urged her husband to act after seeing images of washed-out Midwestern farms on the TV. Powell took up the call for a pipeline to bring "excess" Mississippi River water to the Sonoran Desert and was joined by Tim Dunn, a prominent Republican congressman from Yuma, the miserably hot farming town on the California border that supplies much of the country's winter lettuce.

"Arizona has long been at the forefront among Western states in supporting the development and implementation of pioneering, well-reasoned water management policies," Dunn wrote in an appeal to his fellow legislators. "Diverting [Mississippi River floodwater], which is otherwise lost into the Gulf of Mexico, would also help prevent the loss of human life and billions in economic damages when such flooding occurs."[79]

As imagined, the Mississippi pipeline would run a near straight line skirting the heights of the Rockies from Davenport, Iowa, to Rock Springs, Wyoming. There, it would enter the Colorado watershed through the Green River. Its critics note that traversing about 1,000 miles and gaining more than a mile in elevation would require an enormous amount of energy. The CAP, which covers one third the distance and half the elevation, remains Arizona's foremost power guzzler. Environmentalists also point to myriad risks, which most obviously include the introduction of invasive species and pathogens to the Colorado system. And then the regulatory nightmare: stringent national and state environmental regulations, vehement opposition from local conservation groups, and a compact that forbids the Great Lakes states from sending their water out of the basin without majority consent.

Peter Gleick, a policy expert with the Pacific Institute, a water issue think tank, has said for years that this a harebrained idea destined to go nowhere, and he still believes that. The Mississippi pipeline is technically feasible, he told me, but it will never, ever happen. "It would cost tens or even hundreds of billions of dollars and lead to vast environmental destruction and devastation. Half a century ago, we didn't know about the ecological consequences of massive water diversions, or we didn't care, but those days are over."

Maybe not, in Arizona. Few states have resisted entering the twenty-first century like the Grand Canyon State, where abortion bans date to 1864, shipping containers are dumped along the border to keep out Mexicans, and donkeys may not lawfully sleep in bathtubs. In 2021, in a state that maintains one of the country's lowest high school graduation rates while spending less on public education than all but three of its peers, legislators voted on whether to earmark $160 million for considering the Mississippi pipeline.[80] The resolution passed both houses with strong bipartisan support.

"What a crazy, whacked out bill this is. I just thought we need to fix our own problems," Democratic state senator Victoria Steele told the *Arizona Daily Star* of her initial reaction. But after talking with friends who remembered how the CAP had also seemed like a whacked-out, crazy idea and yet became the very real backbone of the state's subsequent growth, she changed her tune. "The more I talk with experts in hydrology, the more convinced I am that this is an idea that deserves consideration and it's not all that far-fetched," she said. Agricultural communities that are seeing cutbacks in Colorado River water stand the most to gain from a Mississippi pipeline (in the unlikely event that they could afford its water), which is why Miller, Powell, and Dunn support the idea. But Steele represents Tucson, and she's not the only urban representative coming around to the idea.

"It's not just the hearsay out there of going to the Mississippi and such, but trying to figure out how to get water here to this great state that continues to grow and grow. As we get more people we need more water," said David Gowan, a Republican from Sierra Vista.

The idea of this pipeline is not going away, said Raul Grijalva, a US Democratic senator from Tucson. "As this mega drought continues to affect the basin states, including Arizona, there is going to be a look beyond recycling, conservation, beyond 'making do with what we have' strategies."

There have been plenty of whacked-out technocratic solutions to Arizona's aridity: a pipeline to the Great Lakes, an aqueduct to the Columbia River, icebergs towed from Alaska. Gleick has a name for these schemes: "They're zombies," he said. "They're dead, but they still wander around and bother us."

This pipeline idea has remarkable zeal for life. And in light of the attention it continues to garner among top officials, I worry that the costliest mistake we could make when it comes to pipe dreams is not resurrecting dead ones but ignoring those that appear too outlandish to take seriously. After all, the "mad man's" CAP was not the first pipeline proposed for central Arizona. There had been other batty schemes before it. As Jennifer Zuniga of the Bureau of Reclamation explains, one of those, a failed proposal from the 1920s called the Arizona Highline Canal, ultimately paved the way for the CAP, "because the idea would not die."[81]

~

As of the summer of 2022, Arizona had earmarked $1.2 billion to find outside water.[82] The Mississippi pipeline is a favorite among the agriculture community, which has yet to learn its lesson, but most agree that a more realistic option is desalinating water from the Sea of Cortez and piping it the shorter distance to Yuma. Realistic not because delivering desalted water from Mexico is much more feasible than a Mississippi pipeline but because it seems inevitable.

While in office, governor Doug Ducey hoped Arizona would be part of "the largest desalination project in history, anywhere around the globe." His office established ties with Israeli firms that offered to make this dream a reality in Puerto Peñasco, the Mexican beach town where American students on spring break make terrible choices.[83] By Christmas of 2022, Arizona lawmakers were considering an agreement with IDE Technologies to purchase desalted water from the plant the Israeli company plans to build. IDE has constructed desalination facilities all over the world. One of them is the Claude "Bud" Lewis plant in Carlsbad, California, which Gleick had counted among the West's zombie ideas until it came online in 2015. "Carlsbad was a bad idea," Gleick said. "It's still far more expensive than other options for San Diego, but they built it and now they're paying for it."

Reverse osmosis desalting plants like the one in Carlsbad suck up seawater and pour it through filters of sand and anthracite to remove larger contaminants. It is then forced under extremely high pressure through hundreds of canisters inside of which are membranes with infinitesimally small holes that allow water molecules but no salts to pass. What comes out on the other side is so clean that the Carlsbad plant adds minerals before piping it to customers. The challenge is pushing water through such tiny pores, and this is what consumes most of the energy.

On average, desalination uses four times as much power as recycling municipal water, ten times more than treating surface water, and nearly twenty times what it takes to pump groundwater 200 feet.[84] Since most of Arizona's electricity comes from natural gas, the burning of which produces greenhouse gases that trap atmospheric heat, drive climate change, and worsen drought, desalination can be credited with contributing to the problem it seeks to solve. Plants are also vulnerable to power supply disruptions, like if the grid goes down because everyone is cranking their air conditioners and mining Bitcoin. Oceanside facilities risk sucking in thousands of marine creatures, killing the plankton, fish eggs, and larvae that support all life in the marine kingdom. And they create a toxic byproduct, a briny concentrate of all the things we don't want to drink that must be disposed of and is typically blown out to sea, to the detriment of vaquitas, sea turtles, and local fisheries.

Arizona has already tried its hand at desalination. The result stands quietly out near Yuma. In 1993, the Bureau of Reclamation finished the Yuma Desalting Plant to help the United States meet its obligation to send clean water to Mexico.[85] Heavy upstream use and runoff from American farms had left the Colorado River toxically brackish by the time it reached the border. The plant is capable of processing 72 million gallons of that filthy swill into something Mexico and the suffering riverine ecosystem can use. But because of the astronomical cost of operation and 325-ton-per-day output of calcium carbonate and magnesium hydroxide waste sludge, it has never been operated at full capacity. It just sits there, a $245-million testament to shortsighted solutions.

Julie Korak is a civil engineer at the University of Colorado who studies the next generation of desalting technology. Even she has reservations. "Our expectations for desalination are too high," she told me. "Cities should be putting much more effort and investment into reuse instead of going out looking for more water far away." But in America, public perception and funding institutions that prioritize proven solutions have stymied wastewater reuse, an option lamentably labeled "toilet to tap." "If we can get the public OK with reuse, it's hard to justify investing more into seawater reverse osmosis," Korak said.

Gleick optimistically believes that America has outgrown its penchant for massive infrastructural solutions. "There are a lot of engineers out there who think what we need to do is what we've always done: build dams, build plants, take more water out of the ecosystem. That's the old way of doing it, but those days are over, whether they know it or not," he said.

After what I've seen in the Ganges Delta, however, I worry that climate change is just the sort of thing to cause a relapse to the 1950s-era stopgap construction in a place as backward as my home state. Climate change can be catalyst for institutional change, but it can also provide an altruistic cover for self-interested investment. Legislators who laugh off the existence of climate change are hawkish in their search for more water, perhaps because it looks like something they can fix—and fixing it looks good come election time. Politicians are immortalized in bronze busts for constructing dams; no one remembers the heat wave that didn't happen.

Ducey has said he wants a new desalination plant to secure water for Arizona for one hundred years. To see what happens afterward, just look at Miller. He and I met one hundred years after his family arrived in Arizona, and although there had been plenty of good times along the way, a century in the desert has left his family high and dry. Arizona's negotiations with IDE came exactly one hundred years after the signing of the Colorado Compact, which was supposed to solve the West's water woes. The CAP didn't even survive a century before another crisis emerged. One hundred years is a nice round number, but it bears little meaning amid a drought that may last millennia.

In light of all I've seen elsewhere in the world, I read through lofty plans for pipelines and desalination plants and think that this is not adaptation. These are coping mechanisms, technological fixes that coddle consumers and support existing conditions.

Desalination currently requires 2 gallons of seawater to produce 1 gallon of drinking water that costs ten times more than what already comes out of our taps. And yet it remains less legally fraught than taking water from farmers, less politically divisive than requiring drastic conservation, and less imaginative than rethinking our commitment to unmitigated growth. Also, by entering into a binational agreement with Mexico, Arizona can offload some of the most significant challenges—namely the toxic byproduct—to a nation with fewer environmental regulations. Like the Navajo at Black Mesa and the CAP-dependent farmers, Mexican communities would be saddled with the externalities of a boom that omits them. One hundred years on, when the population has outstripped the plant's capabilities and our remaining wild lands have been mined for minerals to produce the miraculous supply of renewable energy needed to maintain our current lifestyles "cleanly," our grandkids will have to figure out what to do next.

The willingness to take on and pass along such a burden must come from some burning necessity. In Tohoku, it was clear: No one could walk among the wreckage of Kesennuma on March 12, 2011, and not believe the coast needed improved protections. In Bangladesh, even where the side effects of embankments were painfully obvious, the chance of saving tens of millions of lives still makes shoring up crumbling levees a strategy worth considering. The motivation is also apparent in Arizona: greed.

There's just so much money to be made on growth. All that's left to do is to succumb to the quintessential American folly of believing that whatever it is, it won't happen here, that some new technology will save the day, that we will somehow evade the worst-case scenario, and that despite a century of evidence to the contrary, history will not repeat itself.

Circles in the Desert

There was once a people who might have seen the cyclical nature of desert living. They left handprints and spirals etched in rock that appear to trace an unending cycle. Whether seasonal or something greater we'll never know, but their messages are out there. We built atop them. At face value, we share little with the Hohokam, named by archeologists for a bastardization of the O'odham word *huhugam*, meaning something like "those who are gone." They lived long ago, had a fraction of our population, and disappeared, which is something none of us can comprehend. But I think focusing on their collapse misses the point.

The Hohokam were the first people in the Western Hemisphere known to have farmed with irrigation. At their peak, they operated 500 miles of canals that watered more than 100,000 acres for roughly 80,000 people.[86] Then, about 1,100 years ago, the good times came to a sudden and mysterious end as a consequence of compounding drought, political strife, and the breakdown of crucial trade networks.

Archaeological sites from that time chronicle a period of conflict when people built defensive structures like the cliff dwellings and lookouts at Bears Ears National Monument in southern Utah. Perhaps those living at higher elevations, who were more dependent on rainfall, suffered first, and maybe their struggles trickled down on the Hohokam in the Phoenix valley. What we know is that the Hohokam built a structured, stable society full of people who didn't think it would ever collapse either. But over time, families began moving away, and their complex canals, which depended on a large village to manage and maintain them, fell into disrepair. Eventually, their pueblos stood empty, lifeless, and baking under the uncaring sun.

Today, there are Indigenous methods for sharing surface water and varieties of crops that have thrived here long before the CAP, but of all the things Anglos can learn from Indigenous experience, the most valuable may be a warning about stretching ourselves so thin that a shock, or a series of them, will become too much to handle. Early settlers ignored this when they chased after growth they could not support locally, and cities are now building on the literal and figurative edge too. What happens when the power and water become so expensive that companies no longer build their headquarters in the desert, employees flee, and property values plummet?

One of the great challenges facing Americans in the Sonoran Desert is our lack of experience. Successful adaptation requires deep knowledge of the place, but our history in the desert begins a mere 150 years ago with infrastructure and has never tried anything else. Hohokam society may have failed, but only after centuries in which they developed a rich culture built out of the landscape. We should take note. When it sprouts from the land, culture is adaptation. From its strip malls and neighborhoods to its golf courses and surf parks, Arizona's modern culture is largely a foreign import that only feigns local roots with adobe facades.

Fraught though it is, the agricultural world maintains one of the only remaining direct links between the people and this place. We can house data anywhere; farmers depend on the land. A subtle side effect of the economic shift to professional and technical industries is the untethering from the desert environment, but in truth this is a continuation of what settlers started a century ago: engineering the landscape so that it would produce for external aims. To this end, the desert's value has always been its malleability. Flat expanses and consistent weather offer a blank slate to farm, given the water, or house industry, given the power. And viewing the land so simply takes us farther from it.

If we want to live in this environment for a long time to come, we will need to maintain contact. This is a role that rural communities are well suited to fulfill, which is why the creep of concrete and the suburbanization of agricultural lands that is now commonly viewed as an adaptation to climate change is actually maladaptive. It severs one of the only practical links we have left.

In a recent gathering of water experts and policymakers in Las Vegas, the urban–rural divide was as wide as ever. Clearly, agriculture's use of Arizona's groundwater and the entire Colorado River system is a problem. In the short term, massive fallowing of farmland would save the aquifers and shore up levels in Lake Mead. Outside central Arizona, where farmers depend on the CAP, many growers still maintain legal seniority on the Colorado. More than ever, though, discussions hinted at how the looming crisis might supersede historic binds.

Kathryn Sorensen, the former water manager for the City of Phoenix, voiced a sentiment few would argue with, saying, "It would be immoral to use water to grow crops while cities go without." Then, when asked about the role of urban population growth in driving the crisis, she said something that caught my ear: "People migrate toward opportunity."

If the history of Arizona has proven anything, it's that opportunity does not materialize out of thin air. What opportunity there is here resulted from aggressive and choreographed efforts made in spite of natural limitations. Farming families like the Millers would not have come to Arizona if not for federal acts to advance westward migration, state incentives to lure settlers, bank loans and boosters' promises of riches, and the heavily subsidized construction of infrastructure like the CAP. Seeing how it has worked out for them, I wonder whether it's not immoral to encourage settlement in a place where generations are locked into living with extreme heat, poor air, and water rationing and where, by arriving, new settlers are forcing others out.

So far, though, it's business as usual. As evidenced by continued talks over new pipelines and desalting plants, engineers and policymakers believe that the same technical logic that caused the climate crisis can be used to fix it. Others know that adaptation on the scale needed requires something larger. It demands a transformational change in how we live. This must happen at the level of culture and the level of choice, like the one about where to live.

Migration stands among the oldest adaptations in human history and continues still. Americans are hardly surprised that millions of Bangladeshis are on the move to escape rising seas or that entire Japanese towns are relocating uphill to avoid tsunamis or that Inuit villages are shifting away from melting coastlines. Despite inhabiting just 3.5 percent of the total area of the planet's third-largest country, we struggle to fathom a reality in which we might have to retreat from a desert that's getting drier. We will eventually accept graywater use requirements, lawn watering restrictions, and other minor reductions in the name of conservation. But, as the Sierra Club's defeat in the 2000s shows, our leaders are unlikely to consider meaningful bounds on unsustainable growth.

Regardless, the Sonoran Desert is going to get very hot for a very long time—migration is inevitable. To alleviate future pain, our adaptive decisions now should consider where we encourage settlement and prioritize places that are most easily sustained. This has never been the desert. Such talk is heresy, I know. Full consumptive use of the region's resources is guaranteed under the Colorado Compact, just as consumer capitalism is baked into American identity. We can more easily imagine an apocalypse than an end to either. And yet, the apocalypse may be happening already.

Standing in the empty streets of Eastmark, I felt as if I had wandered into a dummy town built to test the effects of an atomic blast. The place was so eerily uniform and still. Looking toward Phoenix at dusk, I couldn't help but imagine the whole thing disintegrating in a superheated eruption. The real apocalypse is far less dramatic. It's the slow creep of concrete, the fortressification of quaint hamlets, the entombing of rivers, the unending string of small fixes that build upon each other like mismatched blocks until we've walled ourselves within a world of our own making, cut off from the desert that brought so many of us here. A biosphere that becomes a prison.

What unsettled me most about that street was the void, the yawning hole where a serious concern should be. Nothing about what's happening in Eastmark or any of the other developments or on the massive farms pumping groundwater way out in the desert offer the slightest hint that anything is wrong at all. Given the century of hand-wringing and coffer plundering that's been needed to keep this ship afloat, it's a wonder we don't at least build new homes with the desert in mind. Instead, after all we've constructed, all the land we've blasted and drowned, all the lives we've upended or polluted, the one thing we have yet to adapt is ourselves.

Acknowledgments

This book began as a note scribbled in a margin and grew into an obsession. No one weathered more of that process than my beautiful wife, Emily, and there's no way I would have completed it without her. I am also lucky to have parents who have made all things seem possible and a sister who sets the bar high.

Reporting this book required covering a lot of ground on a shoestring budget and amid a global pandemic. I benefited at every step of the way from the graciousness of old and new friends. Although it's futile to try to list them all, I'd like to say thanks to many who don't appear in the text, especially Laurie Abkemeier, who is so much more than an agent and whose even-keeled belief kept me rowing. Thanks also to Erin Johnson and everyone at Island Press for believing in this project and giving me the opportunity.

Some of the early research was done during my time as a Ted Scripps Fellow at the Center for Environmental Journalism at the University of Colorado. I'd like to thank Erin Ashbaugh for answering every email and helping me cross international borders. And Amanda Carrico, who let me into her classroom and then opened the doors to the world. An enormous thanks to Elizabeth Royte, a friend and mentor of monumental stature. To Peter Brannen for always buying one more round. To Hillary Rosner and Erin Espelie for your helpful critique, and to Terry Wimmer for the brutal honesty. Thanks to the Food and Environmental Reporting Network and to the editors at *National Geographic*, *The Guardian*, and *Huffington Post* for the magazine assignments that almost covered my travel expenses.

Nothing restores faith in humanity like the support of strangers in foreign places. In Japan, I owe an enormous debt to Sébastien Penmellen Boret; I hope we can share that meal one day. A huge thanks to Kumi Aizawa, Nobuo Togashi, and Mayuko Sakamoto; I'll see you on the trail. And to Takaharu Saito and Ulala Tanaka; your knowledge and generosity have inspired me. To Toshiaki Kimura for the folklore and Hiroki Takakura for the history. Thanks also to Hirofumi Endo, Alyne Delaney, Suzuki Wakako, Akiko Iwasaki, Hatekeyama Kohji, Takeshi Komatsu, Waichiryo Katayama, Isabelle Provost, and Jun Sasaki for the scallops, and to Hiroshi Sato, Fumio Sakamoto, Ichiyo Kanno, Michita Onodera, Saito Kazue, Matsumoto Masahiro, Tarumi Owada, Taka Matsumoto, and Takahiro Ito. And to the Nozawa sisters, Suzuna and Kawo, for the unexpected hope.

In Bangladesh, I must first thank my bhā'i, Hafiz Rahaman; I look forward to our next visit to the misti shop. And, of course, Riton Camille Quiah, who always knew where to eat. Thanks also to Anwar Bhuiyan, for getting me across the Padma and opening your home. To all those who shared their knowledge, including Saleemul Huq, Steve Goodbred, Kimberly Rogers, Jeremiah Osborne-Gowey, Kelsea Best, Bishawiit Mallick, Jason Cons, Andrew Jenkins, Steven Rubinyi, Catharien Terwisscha van Scheltinga, Sarder Shafiqul Alam, Abdullah Harun Chowdhury, Kerry Key, Mark Person, Bachchu, and the crew of the *Kokilmoni*. To Kashifa Ahmed and Rofik for welcoming me to Dhaka. To Argina; I hope you are well. To Md. Dulal Akond, S. M. Badal, and Amal Daha in Mongla; Habibura Rahman, Richard Biswas, Nasima Khatun, and Swapan Kumar at Nijera Kori; Topon Mondol, Hobi Moral, and May-hayrun Nessa; and especially Shuchitra Sardar for sharing Karuna's story.

Few subjects are as challenging to write about as home. In Arizona, I'd like to thank the following people for helping to make sense of it all: John Berggren, Paul Lander, Lisa Button, Anne Castle, Susanne Moser, Kathy Jacobs, Kevin Moran, Rimjhim Aggarawal, Ladd Keith, Carol Wessman, Katja Friedrich, Marvin Brown, Brian Betcher, David Dierig, and the bartenders at McMashers in Casa Grande. To the farmers, including Brian Davis, Dan Thelander, Gary Dean, Jerry Turner, Selwyn Justice, Kelley Freeman, and Terry and Ramona Button. To Kyle Woodson and David Dejong for making sense of the past. A big thanks to Sandy Bahr and Peter Else for the grounding. To Rick Brusca, Doug Hendrix, Margaret Wilder, Jamie McAvoy, and Julie Anne Korak for helping me with desalination. To Pat Schoneck, for sharing a story that deserves its own book. And to Ryan Bloom for representing what I love most about the desert.

I would have been dead in the water without the reporters at *The Daily Star* in Dhaka, Tony Davis at *The Arizona Daily Star*, Ian James at *The Arizona Republic*, and Heather Sackett at *Aspen Journalism*, in particular. The world runs on local journalism—we desperately need to fund it.

Finally, thanks to the skaters at the Hartmann Inline Hockey Rink in Boulder, Colorado, for showing up every Thursday and asking how this thing was coming along. And, of course, thanks to the spiders for tolerating me in a corner of the basement. I hope I wasn't a nuisance.

Notes

EPIGRAPH

1. David G. Mccullough, *The Johnstown Flood* (New York: Simon & Schuster Paperbacks, 2018).

INTRODUCTION

1. Jon Barnett and Saffron O'Neill, "Maladaptation." *Global Environmental Change* 20, no. 2 (May 2010): 211–13. https://doi.org/10.1016/j.gloenvcha.2009.11.004.
2. Forest History Society, "U.S. Forest Service Fire Suppression," n.d. https:// foresthistory.org/research-explore/us-forest-service-history/policy-and-law/fire-u-s-forest-service/u-s-forest-service-fire-suppression/#:~:text=In%201935%2C%20the%20Forest%2Service.
3. Anne Barnard, "The $119 Billion Sea Wall That Could Defend New York . . . or Not," *The New York Times*, January 17, 2020, sec. New York. https://www.nytimes.com/2020/01/17/nyregion/the-119-billion-seawall-that-could-defend-new-york-or-not.html.
4. Warren Cornwall, "Shelter from the Storm: Can a Giant Flood Barrier Protect Texas Cities from Hurricanes?," www.science.org, November 17, 2022. https://www.science.org/content/article/shelter-storm-can-giant-flood-barrier-protect-texas-cities-hurricanes.

5. Haley Smith, "California Coastal Commission OKs Desalination Plant in Orange County," *Los Angeles Times*, October 13, 2022. https://www.latimes.com/california/story/2022-10-13/california-coastal-commission-oks-desalination-plant-in-orange-county.
6. "Cloud Seeding Expanding to Boulder County," KUSA.com, 2022. https://www.9news.com/article/weather/cloud-seeding-boulder-county/73-91bf9436-068b-4da4-867c-ab708c6b8bef.

PART I. *SOUTEI-GAI—NORTHEASTERN JAPAN*

1. Rabindranath Tagore. *Personality: Lectures Delivered in America* (New York: Wentworth Press, 2019).
2. Becky Oskin, "Japan Earthquake & Tsunami of 2011: Facts and Information," *Live Science*, September 13, 2017. https://www.livescience.com/39110-japan-2011-earthquake-tsunami-facts.html.
3. Kenneth Chang, "Quake Moves Japan Closer to U.S. and Alters Earth's Spin," *The New York Times*, March 13, 2011. Retrieved March 14, 2011.
4. National Centers for Environmental Information (NCEI), "On This Day: 2011 Tohoku Earthquake and Tsunami," March 11, 2021. https://www.ncei.noaa.gov/news/day-2011-japan-earthquake-and-tsunami#:~:text=The%20March%2011%2C%202011%2C%20earthquake.
5. "The Impact of Casualties of 20,000+: Deaths and Missing Persons by Municipalities," tohokugeo.jp. http://tohokugeo.jp/articles/e-contents1.html.
6. The name has been changed.
7. Thomas Plümper, Alejandro Quiroz Flores, and Eric Neumayer, "The Double-Edged Sword of Learning from Disasters: Mortality in the Tohoku Tsunami," *Global Environmental Change* 44 (May 2017): 49–56. https://doi.org/10.1016/j.gloenvcha.2017.03.002.
8. Giancarlos Troncoso Parady, Bryan Tran, and Stuart Gilmour, "Effect of Seawalls on Tsunami Evacuation Departure in the 2011 Great East Japan Earthquake," *Injury Prevention* 25, no. 6 (November 17, 2018): 535–39. https://doi.org/10.1136/injuryprev-2018-042954.

9. Sébastien Penmellen Boret and Julia Gerster, "Social Lives of Tsunami Walls in Japan: Concrete Culture, Social Innovation and Coastal Communities," *IOP Conference Series: Earth and Environmental Science* 630 (January 13, 2021): 012029. https://doi.org/10.1088/1755-1315/630/1/012029.
10. "Japan's Tsunami Defences Brutally Exposed," web.archive.org, March 4, 2016. https://web.archive.org/web/20160304032001/http://news.asiaone.com/News/Latest+News/Asia/Story/A1Story20110326-270285.html.
11. Natsuko Fukue, "The Towering Sea Wall Legacy of Japan's 2011 Tsunami," The Jakarta Post, March 5, 2021. https://www.thejakartapost.com/life/2021/03/05/the-towering-sea-wall-legacy-of-japans-2011-tsunami.html.
12. Roshanak Nateghi, Jeremy D. Bricker, Seth D. Guikema, and Akane Bessho, "Statistical Analysis of the Effectiveness of Seawalls and Coastal Forests in Mitigating Tsunami Impacts in Iwate and Miyagi Prefectures," edited by Yoshiaki Taniyama, *PLoS One* 11, no. 8 (August 10, 2016): e0158375. https://doi.org/10.1371/journal.pone.0158375.
13. Statistics Bureau, Ministry of Internal Affairs and Communications, "Statistics Bureau Home Page/Statistical Handbook of Japan 2020," Stat.go.jp, 2020. https://www.stat.go.jp/english/data/handbook/c0117.html#:~:text=Japan.
14. "The Impact of Casualties of 20,000+: Deaths and Missing Persons by Municipalities," tohokugeo.jp. http://tohokugeo.jp/articles/e-contents1.html.
15. Shuhei Kimura, "When a Seawall Is Visible: Infrastructure and Obstruction in Post-Tsunami Reconstruction in Japan," *Science as Culture* 25, no. 1 (January 2, 2016): 23–43. https://doi.org/10.1080/09505431.2015.1081501.
16. "Symposium Revisits 17th Century Earthquake and Tsunami," www.tohoku.ac.jp, January 4, 2022. https://www.tohoku.ac.jp/en/news/research/symposium_revisits_17_century_earthquake_and_tsunami.html.
17. Brian F. Atwater, *The Orphan Tsunami of 1700: Japanese Clues to a Parent Earthquake in North America* (Reston, Va.: US Geological Survey, 2005).

18. "The Great East Japan Earthquake Restoration Report 2011–2013," Miyagi Prefectural Government, 2015.
19. Anawat Suppasri, "Lessons Learned from the 2011 Great East Japan Tsunami: Performance of Tsunami Countermeasures, Coastal Buildings, and Tsunami Evacuation in Japan," *Pure and Applied Geophysics* 170, no. 6–8 (July 7, 2012): 993–1018. https://doi.org/10.1007/s00024-012-0511-7.
20. National Centers for Environmental Information (NCEI), "On This Day: 2011 Tohoku Earthquake and Tsunami," March 11, 2021. https://www.ncei.noaa.gov/news/day-2011-japan-earthquake-and-tsunami#:~:text=The%20March%2011%2C%202011%2C%20earthquake.
21. Anawat Suppasri, Panon Latcharote, Jeremy D. Bricker, Natt Leelawat, Akihiro Hayashi, Kei Yamashita, Fumiyasu Makinoshima, Volker Roeber, and Fumihiko Imamura, "Improvement of Tsunami Countermeasures Based on Lessons from the 2011 Great East Japan Earthquake and Tsunami—Situation after Five Years," *Coastal Engineering Journal* 58, no. 4 (December 2016): 1640011-1–1640011-30. https://doi.org/10.1142/s0578563416400118.
22. Jan Oetjen, Vallam Sundar, Sriram Venkatachalam, Klaus Reicherter, Max Engel, Holger Schüttrumpf, and Sannasi Annamalaisamy Sannasiraj, "A Comprehensive Review on Structural Tsunami Countermeasures," *Natural Hazards*, May 16, 2022. https://doi.org/10.1007/s11069-022-05367-y.
23. P. Matanle, J. Littler, and O. Slay, "Imagining Disasters in the Era of Climate Change: Is Japan's Seawall a New Maginot Line?," edited by M. Selden, *The Asia-Pacific Journal: Japan Focus* 17, no. 13 (July 1, 2019): 1–29. http://eprints.whiterose.ac.uk/148388/.
24. Nathan Hopson, "Systems of Irresponsibility and Japan's Internal Colony," *The Asia-Pacific Journal: Japan Focus*, December 27, 2013. https://apjjf.org/2013/11/52/Nathan-Hopson/4053/article.html.
25. Ichiro Numazaki, "Too Wide, Too Big, Too Complicated to Comprehend: A Personal Reflection on the Disaster That Started on March 11, 2011," *Asian Anthropology* 11, no. 1 (January 2012): 27–38. https://doi.org/10.1080/1683478x.2012.10600853.

26. Michael Fisch, "Japan's Extreme Infrastructure: Fortress-ification, Resilience, and Extreme Nature," *Social Science Japan Journal*, June 30, 2022. https://doi.org/10.1093/ssjj/jyac011.
27. Kimura, "When a Seawall Is Visible."
28. Nateghi et al., "Statistical Analysis of the Effectiveness of Seawalls."
29. "Ogatsu School Restoration Project," ogatsusaisei.jp. http://ogatsusaisei.jp/english/.
30. "Japan Cement Production—October 2022 Data—1985–2021 Historical—November Forecast," tradingeconomics.com. https://tradingeconomics.com/japan/cement-production.
31. Michael Fisch, "Concrete Sovereignty," *Machinicnatures*. http://www.machinicnatures.net/fieldwork-1.
32. Gavan McCormack and Norma Field, *The Emptiness of Japanese Affluence* (London: Routledge, 2016).
33. "Concrete Needs to Lose Its Colossal Carbon Footprint," *Nature* 597, no. 7878 (September 28, 2021): 593–94. https://doi.org/10.1038/d41586-021-02612-5.
34. Man-Shi Low, "Material Flow Analysis of Concrete in the United States," Massachusetts Institute of Technology, 2005. http://dspace.mit.edu/handle/1721.1/33030#:~:text=Abstract.
35. M. Garside, "Cement Production Global 2021," *Statista*, April 1, 2022. https://www.statista.com/statistics/1087115/global-cement-production-volume/#statisticContainer.
36. Numazaki, "Too Wide, Too Big, Too Complicated to Comprehend."
37. Dinil Pushpalal, "The Great Eastern Japan Earthquake 11 March 2011—Lessons Learned and Research Questions," 2013. https://www.preventionweb.net/files/42625_42625thegreateasternjapanearthquake.pdf.
38. Isabel Galwey, "Taking Root: Failure and Success for the Great Forest Wall of Tōhoku," *Anthroposphere*, August 15, 2020. https://www.anthroposphere.co.uk/post/taking-root-failure-and-success-for-the-great-forest-wall-of-t%C5%8Dhoku.
39. "Fishing Vessel Carried Inland by Tsunami to Be Dismantled," *The Japan Times*, March 26, 2013. https://www.japantimes.co.jp/news/2013/03/26/national/fishing-vessel-carried-inland-by-tsunami-to-be-dis mantled/.

40. Y. S. Yadava and Masaaki Sato, "Fishing Ports of Japan," *Bay of Bengal News*, 2008. https://bobpigo.org/html_site/bbn/sep-dec08/sep-dec2008_pages41-45.pdf.
41. "Ruins of the Great East Japan Earthquake, Kesennuma City Memorial Museum," www.kesennuma-memorial.jp. https://www.kesennuma-memorial.jp/english/#:~:text=in%20your%20browser.-.
42. Satoru Imakawa, "Finally Fully Released. Tsunami Re-simulation Results [Kesennuma City]," Kesennuma City Council Member Satoru Imagawa's activity report, June 4, 2022. https://imakawa.net/blog/6668.html.
43. "March 11, 2011 Great East Japan Earthquake—A Huge 'Black' Tsunami Hit Miyako City [Mainichi Bosai]/Great East Japan Earthquake, Tsunami," November 3, 2011. https://www.youtube.com/watch?v=4XvFFfgXwnw&t=181s.
44. Hajime Chiba, "Living with the Sea: The Folklore of Adaptive Reconstruction," 2017, p. 164. https://www.iubs.org/fileadmin/user_upload/Biology-International/BI-Specials/BI_Special_Issue_No-36_beta_2_web.pdf.
45. Takeshi Yuhara, Takao Suzuki, Tatsuki Nishita, Junichi Murakami, Wataru Makino, Gen Kanaya, Kyoko Kinoshita, Natsuru Yasuno, Takashi Uchino, and Jotaro Urabe, "Recovery of Macrobenthic Communities in Tidal Flats Following the Great East Japan Earthquake," *Limnology and Oceanography Letters*, November 10, 2022. https://doi.org/10.1002/lol2.10292.
46. Harufumi Nishida, Jun Yokoyama, Steven Wagstaff, and Paul Callomon, "DAB: Disaster and Biodiversity," IUBS Biology International Special Issue no. 36," July 31, 2017. https://iubs.org/wp-content/uploads/2021/12/BI_Special_Issue_No-36_beta_2_web.pdf.
47. Nima Hosseinzadeh, Mohammad Ghiasian, Esber Andiroglu, Joel Lamere, Landolf Rhode-Barbarigos, James Sobczak, Kathleen Sullivan Sealey, and Prannoy Suraneni, "Concrete Seawalls: Load Considerations, Ecological Performance, Durability, and Recent Innovations," July 20, 2021. https://doi.org/10.31224/osf.io/h6zt8.

48. Eduardo Jaramillo, Jenifer Dugan, David Hubbard, Mario Manzano, and Cristian Duarte, "Ranking the Ecological Effects of Coastal Armoring on Mobile Macroinvertebrates across Intertidal Zones on Sandy Beaches," *Science of the Total Environment* 755 (February 2021): 142 573. https://doi.org/10.1016/j.scitotenv.2020.142573.
49. Timothy S. Lee, Jason D. Toft, Jeffery R. Cordell, Megan N. Dethier, Jeffrey W. Adams, and Ryan P. Kelly, "Quantifying the Effectiveness of Shoreline Armoring Removal on Coastal Biota of Puget Sound," *PeerJ* 6 (February 23, 2018): e4275. https://doi.org/10.7717/peerj.4275.
50. From interview with Akihiko Sugawara, founder of Kesennuma City Sea Wall Study Group, chairman of Kesennuma Chamber of Commerce & Industry.
51. Michael Carlowicz, "Ten Years after the Tsunami: One of the Hardest Hit Coastal Cities in Japan Is Still Working to Recover," *SciTech Daily*, March 12, 2021. https://scitechdaily.com/ten-years-after-the-tsunami-one-of-the-hardest-hit-coastal-cities-in-japan-is-still-working-to-recover/.
52. Shoma Okamoto, "Sakura Line 311: A Project to Plant Cherry Trees above the Tsunami's Arrival Point in Rikuzentakata City to Pass On the Disaster to Future Generations," July 4, 2012. https://www.sakura-line311.org/english-information.
53. "Sanriku Fukko National Park," www.japan.travel. https://www.japan.travel/national-parks/parks/sanriku-fukko/.
54. "Millennium Hope Hills English Toppage," sennen-kibouno-oka.com. https://sennen-kibouno-oka.com/english/.
55. Suppasri, "Lessons Learned."
56. Matanle, "Imagining Disasters."
57. Noriyuki Shikata, "Press Briefing at the Prime Minister's Office for Members of the Foreign Press," April 14, 2011. https://japan.kantei.go.jp/incident/pdf/foreign-press-briefing-20110414-script-e.pdf.
58. Ted Shaffrey, "Amid Rising Seas, Atlantic City Has No Plans for Retreat," *AP News*, October 12, 2022. https://apnews.com/article/floods-science-travel-climate-and-environment-new-jersey-c3e10d275df462ecfc35b1cde23265c2.

PART II. *PAGAL*, BY ANY OTHER NAME—*SOUTHWEST BANGLADESH*

1. John Steinbeck, *In Dubious Battle/John Steinbeck* (New York: Viking, 1936).
2. "Bangladesh: Displaced and Dispossessed." *Internal Displacement in South Asia*, 2005. https://pdf.zlibcdn.com/dtoken/061b5518fc64d28cff1872b84274ff62/Internal_Displacement_in_South_Asia_The_Relevance_(z-lib.org).pdf.
3. Kasia Paprocki, "Threatening Dystopias: Development and Adaptation Regimes in Bangladesh," *Annals of the American Association of Geographers* 108, no. 4 (January 23, 2018): 955–73. https://doi.org/10.1080/24694452.2017.1406330.
4. "Karunamoyee Sarder Murder Case: 12 Get Life in Khulna," *The Daily Star*, March 23, 2007. http://archive.thedailystar.net/2007/03/23/d70323100197.htm.
5. "BNP Leader Wazed Ali Dead," *The Daily Star*, December 25, 2003. http://archive.thedailystar.net/2003/12/25/d31225100588.htm.
6. "Bangladesh: Impacts of the Shrimp Farming Industry on Women | World Rainforest Movement," www.wrm.org.uy, July 30, 2013. https://www.wrm.org.uy/bulletin-articles/bangladesh-impacts-of-the-shrimp-farming-industry-on-women#:~:text=Around%2055%25%20of%20shrimp%20raised.
7. Marco Keurentjes, "The Original Meaning of the Dutch Word *Polder*," *Amsterdamer Beiträge Zur Älteren Germanistik* 82, no. 3 (October 26, 2022): 319–38. https://doi.org/10.1163/18756719-12340261.
8. Matt Rosenberg, "How the Netherlands Reclaimed Land from the Sea," ThoughtCo, 2019. https://www.thoughtco.com/polders-and-dikes-of-the-netherlands-1435535.
9. "World Rice Production by Country," AtlasBig, January 1, 1970. https://www.atlasbig.com/en-us/countries-by-rice-production.
10. Warren Cornwall, "As Sea Levels Rise, Bangladeshi Islanders Must Decide between Keeping the Water Out—or Letting It In," www.science.org, March 1, 2018. https://www.science.org/content/article/sea-levels-rise-bangladeshi-islanders-must-decide-between-keeping-water-out-or-letting.

11. "Country Climate and Development Report: Bangladesh," The World Bank Group, October 2022. https://openknowledge.worldbank.org/bitstream/handle/10986/38181/CCDR-Bangladesh-MainReport.pdf.
12. UN Environment, "Bangladesh Uncovers the Crippling Cost of Climate Change Adaptation," October 4, 2017. https://www.unep.org/news-and-stories/press-release/bangladesh-uncovers-crippling-cost-climate-change-adaptation.
13. "The Polder Promise: Unleashing the Productive Potential in Southern Bangladesh," CGIAR Research Program on Water, Land and Ecosystem, 2015. https://wle.cgiar.org/sites/default/files/documents/The%20polder%20promise_WLE.pdf.
14. Jeroen F. Warner, Martijn F. van Staveren, and Jan van Tatenhove, "Cutting Dikes, Cutting Ties? Reintroducing Flood Dynamics in Coastal Polders in Bangladesh and the Netherlands," *International Journal of Disaster Risk Reduction* 32 (December 2018): 106–12. https://doi.org/10.1016/j.ijdrr.2018.03.020.
15. Deryck O. Lodrick and Nafis Ahmad, "Ganges River | History, Location, Map, & Facts," in *Encyclopædia Britannica*, May 24, 2019. https://www.britannica.com/place/Ganges-River.
16. Sameer Yasir and Jeffrey Gettleman, "Super Cyclone Bears Down on India and Bangladesh," *The New York Times*, May 19, 2020, sec. World. https://www.nytimes.com/2020/05/19/world/asia/cyclone-amphan-bangladesh-india.html.
17. M. Khalequzzaman, "Assessment of the Bangladesh Delta Plan 2100 and Recommendation to Enhance Its Implementation Strategies," in *Delta Plan 2100 and Sustainable Development in Bangladesh*, ed. M. S. Islam and M. Khalequzzaman (Dhaka, Bangladesh: BAPA-BEN, 2019).
18. Md Ruknul Ferdous, Giuliano Di Baldassarre, Luigia Brandimarte, and Anna Wesselink, "The Interplay between Structural Flood Protection, Population Density, and Flood Mortality along the Jamuna River, Bangladesh," *Regional Environmental Change* 20, no. 1 (February 3, 2020). https://doi.org/10.1007/s10113-020-01600-1.

19. Mohammed Sarfaraz Gani Adnan, Anisul Haque, and Jim W. Hall, "Have Coastal Embankments Reduced Flooding in Bangladesh?," *Science of the Total Environment* 682 (September 2019): 405–16. https://doi.org/10.1016/j.scitotenv.2019.05.048.
20. "The Cost of Corruption in Construction," Royal Institution of Chartered Surveyors, November 24, 2021. https://www.rics.org/north-america/wbef/megatrends/markets-geopolitics/the-cost-of-corruption-in-construction/.
21. "Governance Challenges in Disaster Response and Way Forward: Cyclone Amphan and Recent Experiences," Transparency International Bangladesh, December 24, 2020. https://www.ti-bangladesh.org/beta3/images/2020/report/Amphan/Amphan_Study_ES_Eng.pdf.
22. "Bengal's Irrigation Problems: Sir Willcock's Views Suggested Construction of a Big Barrage," *The Times of India* (1861–2010), March 22, 1928; ProQuest Historical Newspapers: *The Times of India*.
23. "Ganges River," *Geology Page*, January 14, 2015. https://www.geologypage.com/2015/01/ganges-river.html.
24. Lodrick and Ahmad, "Ganges River."
25. C. Wilson, S. Goodbred, C. Small, J. Gilligan, S. Sams, B. Mallick, and R. Hale. "Widespread Infilling of Tidal Channels and Navigable Waterways in the Human-Modified Tidal Deltaplain of Southwest Bangladesh," *Elementa: Science of the Anthropocene* 5 (January 1, 2017): 28. https://doi.org/10.1525/elementa.263.
26. Ibid.
27. Michael S. Steckler, Bar Oryan, Md. Hasnat Jaman, Dhiman R. Mondal, Céline Grall, Carol A. Wilson, S. Humayun Akhter, Scott DeWolf, and Steven L. Goodbred, "Recent Measurements of Subsidence in the Ganges-Brahmaputra Delta, Bangladesh," March 4, 2021. https://doi.org/10.5194/egusphere-egu21-6562.
28. "River Deltas—Concepts, Models, and Examples," SEPM Special Publication no. 83, 2005 Society for Sedimentary Geology, pp. 413–34.
29. L. W. Auerbach, S. L. Goodbred Jr., D. R. Mondal, C. A. Wilson, K. R. Ahmed, K. Roy, M. S. Steckler, C. Small, J. M. Gilligan, and B. A. Ackerly, "Flood Risk of Natural and Embanked Landscapes on the

Ganges–Brahmaputra Tidal Delta Plain," *Nature Climate Change* 5, no. 2 (January 5, 2015): 153–57. https://doi.org/10.1038/nclimate2472.
30. Nitish Sengupta, *Land of Two Rivers* (London: Penguin UK, 2011).
31. Kalyan Rudra, *Rivers of the Ganga-Brahmaputra-Meghna Delta. Geography of the Physical Environment* (Cham, Switzerland: Springer International Publishing, 2018). https://doi.org/10.1007/978-3-319-76544-0.
32. Sengupta, *Land of Two Rivers.*
33. Rudra, *Rivers of the Ganga-Brahmaputra-Meghna Delta.*
34. Camelia Dewan, *Misreading the Bengal Delta : Climate Change, Development, and Livelihoods in Coastal Bangladesh* (Seattle: University of Washington Press, 2021).
35. Ibid.
36. Sengupta, *Land of Two Rivers.*
37. Karl Marx, "The British Rule in India by Karl Marx," www.marxists.org, n.d. https://www.marxists.org/archive/marx/works/1853/06/25.htm.
38. C.E.A.W.O., "Review of *Lectures on the Ancient System of Irrigation in Bengal, and Its Application to Modern Problems* by William Willcocks," *The Geographical Journal* 77, no. 4 (April 1931): 374. https://doi.org/10.2307/1784300.
39. Laxman D. Satya, "British Imperial Railways in Nineteenth Century South Asia," *Economic and Political Weekly* 43, no. 47 (June 5, 2015): 7–8. https://www.epw.in/journal/2008/47/special-articles/british-imperial-railways-nineteenth-century-south-asia.html.
40. S. Watts, "British Development Policies and Malaria in India 1897–c. 1929," *Past & Present* 165, no. 1 (November 1, 1999): 141–81. https://doi.org/10.1093/past/165.1.141.
41. Henry E. Armstrong, "Life and Experiences of a Bengali Chemist," *Nature* 131, no. 3315 (May 1933): 672–74. https://doi.org/10.1038/131672a0.
42. A. M. Rosenthal, "Capital of India a Flood Refuge; Aged Villagers and Cattle Seek Safety in New Delhi—Men Fight the Rivers," *New York Times*, October 9, 1995. https://timesmachine.nytimes.com/timesmachine/1955/10/09/91371088.pdf?pdf_redirect=true&ip=0.

43. "Tidal Channel—Banglapedia," en.banglapedia.org. https://en.bangla pedia.org/index.php/Tidal_Channel.
44. Kimberley Anh Thomas, "The Problem with Solutions: Development Failures in Bangladesh and the Interests They Obscure," *Annals of the American Association of Geographers* 110, no. 5 (February 10, 2020): 1631–51. https://doi.org/10.1080/24694452.2019.1707641.
45. Leendert de Die, "Tidal River Management: Temporary Depoldering to Mitigate Drainage Congestion in the Southwest Delta of Bangla desh," Water Resources Management Group, Wageningen University, March 2013. https://edepot.wur.nl/258920.
46. Sheikh Hasina. *Secret Documents of Intelligence Branch on Father of the Nation* (Bangladesh: Bangabandhu Sheikh Mujibur Rahman. Routledge, 2020).
47. Thomas, "The Problem with Solutions."
48. Q. M. Zaman, "Rivers of Life: Living with Floods in Bangladesh," *Asian Survey* 33, no. 10 (1993): 985–96. https://doi.org/10.2307/2645097.
49. Harun Rasid, "Impact Assessments from Survey of Floodplain Residents: The Case of the Dhaka–Narayanganj–Demra (DND) Project, Bangladesh," *Impact Assessment* 14, no. 2 (June 1996): 115–32. https://doi.org/10.1080/07349165.1996.9725893.
50. Thomas, "The Problem with Solutions."
51. Imtiaz Ahmed, "Governance and Flood: Critical Reflections on the 1998 Deluge," *Futures* 33, no. 8–9 (October 2001): 803–15. https://doi.org/10.1016/s0016-3287(01)00020-9.
52. De Die, "Tidal River Management."
53. "United Nations Appoints Dr Nazrul Islam as Chief of Development Research," *The Daily Star*, August 19, 2021. https://www.thedailystar.net/news/bangladesh/news/united-nations-appoints-dr-nazrul-islam-chief-development-research-2156236.
54. Atiur Rahman, *Beel Dakatia* (Dhaka, Bangladesh: University Press, 1995).
55. Interview with Andrew Jenkins.

56. "Bangladesh —Place Explorer—Data Commons," datacommons.org, n.d. https://datacommons.org/place/country/BGD?utm_medium=explore&mprop=count&popt=Person&hl=en.
57. De Die, "Tidal River Management."
58. "Padma Bridge Will Boost Sacrificial Animal Supply for Eid-Ul-Azha," *Bangla Insider*, April 30, 2022. https://www.banglainsider.com/en/inside-bangladesh/113932/Padma-Bridge-will-boost-sacrificial-animal-supply-for-Eid-ul-Azha.
59. "Padma Bridge So Far Largest Bridge Chinese Companies Built Outside China," *The Daily Star*, June 20, 2022. https://www.thedailystar.net/news/bangladesh/diplomacy/news/padma-bridge-so-far-largest-bridge-chinese-companies-built-outside-china-3051916.
60. "Community People to Become Sundarbans Dolphin Saviours," United Nations Development Programme, January 6, 2019. https://www.undp.org/bangladesh/news/community-people-become-sundarbans-dolphin-saviours.
61. "Padma Bridge Will Open to Traffic by 2022 End: PM," *The Daily Star*, April 6, 2022. https://www.thedailystar.net/news/bangladesh/development/news/padma-bridge-will-open-traffic-2022-end-pm-299 9236.
62. "'A Drop in the Pond,' and Other Lame Justifications—The Case for Rampal Is Sinking Fast," BankTrack, 2016. https://www.banktrack.org/blog/a_drop_in_the_pond_and_other_lame_justifications_the_case_for_rampal_is_sinking_fast.
63. "Five Missing after Ship Sank with About 600 Tons of Coal in Pasur River of Sundarbans," *Marine Insight*, November 16, 2021. https://www.marineinsight.com/shipping-news/five-missing-after-ship-sank-with-about-600-tons-of-coal-in-pasur-river-of-sundarbans/.
64. Abdullah Harun Chowdhury, "Impacts of Industrialization and Infra structure Developments on the Flora, Fauna and Ecosystems of the Sundarbans and Surrounding Areas," *African Journal of Biological Sciences* 4, no. 3 (May 7, 2022): 22–40. https://doi.org/10.33472/afjbs.4.3.2022.22-40.

65. Muhammad Shifuddin Mahmud, Dik Roth, and Jeroen Warner, "Rethinking 'Development': Land Dispossession for the Rampal Power Plant in Bangladesh," *Land Use Policy* 94 (May 2020): 104492. https://doi.org/10.1016/j.landusepol.2020.104492.
66. Alyssa Ayres, "Bangladesh: Capitalist Haven," *Forbes*, October 28, 2014. https://www.forbes.com/sites/alyssaayres/2014/10/28/bangladesh-capitalist-haven/?sh=215168ca145e.
67. "Living with Water: Climate Adaptation in the World's Deltas," Global Center on Adaptation, 2021. https://gca.org/wp-content/uploads/2021/01/Living-with-water-climate-adaptation-in-the-worlds-deltas.pdf.
68. A. K. Magnan, E. L. F. Schipper, M. Burkett, S. Bharwani, I. Burton, S. Eriksen, F. Gemenne, J. Schaar, and G. Ziervogel, "Addressing the Risk of Maladaptation to Climate Change," *Wiley Interdisciplinary Reviews: Climate Change* 7, no. 5 (May 17, 2016): 646–65. https://doi.org/10.1002/wcc.409.

PART III. THE AUDACITY OF DESERT LIVING—*CENTRAL ARIZONA*

1. This poem appears at the Meiji Jingu temple in Tokyo.
2. An acre-foot is enough water to cover an acre 1 foot deep, or 325,851 gallons, which is just less than half the capacity of an Olympic-sized swimming pool.
3. Joanna Allhands, "How Many Farms Can Arizona and California Lose before We Feel It at the Grocery Store?," *The Arizona Republic*, August 5, 2022. https://www.azcentral.com/story/opinion/op-ed/joannaallhands/2022/08/05/arizona-must-rethink-farming-save-massive-water-cuts-loom/10219396002/.
4. "The Law of the Colorado River and Preservation of Imperial Irrigation District's Present Perfected Water Rights," Imperial Irrigation District, January 30, 2018. https://www.iid.com/home/showpublisheddocument/16799/636528130201570000.
5. Lieut. J. C. Ives, Report Upon the Colorado River of the West, Senate Ex. Doc. 36th Congress, 1st Session, p. 110, 1861.

6. John Wesley Powell, *Report on the Lands of the Arid Region of the United States: With a More Detailed Account of the Lands of Utah*, ed. Wallace Stegner (Cambridge, Mass.: Belknap Press, 1962).
7. Thomas R. Van Devender, "The Deep History of the Sonoran Desert," www.desertmuseum.org, n.d. https://www.desertmuseum.org/books/nhsd_deep_history.php.
8. *Colorado River Basin Water Management: Evaluating and Adjusting to Hydroclimatic Variability* (Washington, D.C.: The National Academies Press, 2007). https://doi.org/10.17226/11857.
9. "Geology—Grand Canyon National Park (U.S. National Park Service)," National Park Service, October 17, 2021. https://www.nps.gov/grca/learn/nature/grca-geology.htm#:~:text=Finally%2C%20beginning%20just%205%2D6.
10. "A Look at the Sonoran Desert," Sonoran Desert Museum, 1999. https://www.desertmuseum.org/center/edu/docs/6-8_TIP_background.pdf.
11. Ashley Kerna Bickel, Dari Duval, and George Frisvold, "Contribution of On-Farm Agriculture and Agribusiness to the Pinal County Economy an Economic Contribution Analysis for 2016," Department of Agricultural & Resource Economics, University of Arizona Cooperative Extension, 2018. https://economics.arizona.edu/file/742/download?token=eEwu2-0c.
12. USDA NASS, 2017 Census of Agriculture.
13. Kerna Bickel et al., "Contribution of On-Farm Agriculture and Agribusiness."
14. Erik-Anders Shapiro, "Cotton in Arizona: A Historical Geography," Repository.arizona.edu, 1989. http://hdl.handle.net/10150/291975.
15. "Federally Recognized Tribes in Arizona," Arizona State Museum, n.d. https://statemuseum.arizona.edu/programs/american-indian-relations/tribes-arizona.
16. David H. DeJong, "Forced to Abandon Their Farms: Water Deprivation and Starvation among the Gila River Pima, 1892–1904," *American Indian Culture and Research Journal* 28, no. 3 (January 2004): 29–56. https://doi.org/10.17953/aicr.28.3.l460q665503415x0.

17. Shapiro, "Cotton in Arizona."
18. Ibid.
19. From information held at Gilbert Historical Society, accessed October 28, 2021.
20. Shapiro, "Cotton in Arizona."
21. Census of Agriculture: 1925—Arizona. https://agcensus.library.cornell.edu/wp-content/uploads/1925-Arizona-State_Tables-Table-02.pdf.
22. "Roosevelt Dam—National Historic Landmarks," U.S. National Park Service, August 29, 2018. https://www.nps.gov/subjects/nationalhistoriclandmarks/roosevelt-dam.htm#:~:text=Completed%20at%20a%20cost%20of.
23. Eric Kuhn and John Fleck, *Science Be Dammed: How Ignoring Inconvenient Science Drained the Colorado River* (Tucson: The University of Arizona Press, 2019).
24. Tony Davis, "U.S. West's Water Management System on Edge of Collapse, Expert Says," *Arizona Daily Star*, December 1, 2022. https://tucson.com/news/local/subscriber/u-s-wests-water-management-system-on-edge-of-collapse-expert-says/article_a54ec2ee-6f51-11ed-a244-47832c41e61b.html.
25. Wendell Berry, *The Gift of Good Land: Further Essays Cultural and Agricultural* (Berkeley, Calif.: Counterpoint, 2009).
26. Janick Artiola and Kristine Uhlman, "Arizona Well Owner's Guide to Water Supply," University of Arizona, 2009. https://clu-in.org/conf/tio/srpwir4_072116/Arizona-Well-Owners-Guide-UAz.pdf.
27. Marc Reisner, *Cadillac Desert: The American West and Its Disappearing Water* (New York: Penguin Books, 1986).
28. "Tucson Post World War II Residential Subdivision Development 1945–1973," City of Tucson Urban Planning and Design Department Historic and Cultural Resources, October 2007. https://www.tucsonaz.gov/files/preservation/Text_-_Tucson_Post_WWII_Residential_Subdivision_Development.pdf.
29. AMWUA. "Groundwater in Arizona: Past, Present, and Future—Part One," n.d. https://www.amwua.org/blog/groundwater-what-is-it-and-why-does-it-matter.

30. Michael C. Carpenter, "Earth Fissures and Subsidence Complicate Development of Desert Water Resources," U.S. Geological Survey, 2009. https://pubs.usgs.gov/circ/circ1182/pdf/09Arizona.pdf.
31. "Tucson Post World War II Residential Subdivision Development, 1945–1973," City of Tucson, October 2007. https://www.tucsonaz.gov/files/preservation/Text_-_Tucson_Post_WWII_Residential_Subdivision_Development.pdf.
32. Jennifer E. Zuniga, "The Central Arizona Project," Bureau of Reclamation, 2000. https://www.usbr.gov/projects/pdf.php?id=94.
33. Andrew Curley, "Infrastructures as Colonial Beachheads: The Central Arizona Project and the Taking of Navajo Resources," *Environment and Planning D: Society and Space*, February 20, 2021, 026377582199153. https://doi.org/10.1177/0263775821991537.
34. T. R. Witcher, "The Storied History of the Central Arizona Project," www.asce.org, March 1, 2022. https://www.asce.org/publications-and-news/civil-engineering-source/civil-engineering-magazine/issues/magazine-issue/article/2022/03/the-storied-history-of-the-central-arizona-project#:~:text=As%20planned%2C%20the%20CAP%20delivers.
35. Murielle Coeurdray, Joan Cortinas, Brian O'Neill, and Franck Poupeau, "Sharing the Colorado River: The Policy Coalitions of the Central Arizona Project (Part 2) Introduction: Arizona versus California," 2016.
36. "Engineering 'Marvel' CAP Is a 35-Year-Old Economic Powerhouse," Victoria Harker-Chamber Business News, *AZ Big Media*, June 11, 2020. https://azbigmedia.com/business/engineering-marvel-cap-is-a-35-year-old-economic-powerhouse/.
37. L. Kolankiewicz and R. Beck, "Weighing Sprawl Factors in Large US Cities—Analysis of US Bureau of the Census Data on the 100 Largest Urbanized Areas of the United States," 2001. http://sprawlcity.com/studyUSA/.
38. Robert A. Young and William E. Martin, "The Economics of Arizona's Water Problcms," *Arizona Review* 16(1967): 9-1.
39. H. Ingram, W. E. Martin, and N. K. Laney, "A Willingness to Play: Analysis of Water Resources Development in Arizona," Escholarship.org, 1983. https://escholarship.org/uc/item/0906f2m3.

40. Zuniga, "The Central Arizona Project."
41. Dennis O'Brien, "A New Option for Texas: Pima Cotton," USDA Agricultural Research Service, 2020. https://www.ars.usda.gov/news-events/news/research-news/2020/a-new-option-for-texas-pima-cotton/.
42. Tony Davis, "UA Professor Who Foresaw CAP Farmers' Economic Troubles Dies at 87," *Arizona Daily Star*, September 28, 2020. https://tucson.com/news/local/ua-professor-who-foresaw-cap-farmers-economic-troubles-dies-at-87/article_b549c8ac-cf8c-5864-b5ed-4a22c9417298.html.
43. Amaia Albizua and Alejandra Zaga-Mendez, "Changes in Institutional and Social–Ecological System Robustness due to the Adoption of Large-Scale Irrigation Technology in Navarre (Spain)," *Environmental Policy and Governance*, February 3, 2020. https://doi.org/10.1002/eet.1882.
44. "Colorado River 2007 Interim Guidelines and Drought Contingency Plans." Water Education Foundation, 2023. https://www.watereducation.org/aquapedia/colorado-river-seven-states-agreement.
45. "USGS Groundwater for Arizona: Water Levels," nwis.waterdata.usgs.gov. https://nwis.waterdata.usgs.gov/az/nwis/gwlevels.
46. Ella Nilsen, "Wells Are Running Dry in Drought-Weary Southwest as Foreign-Owned Farms Guzzle Water to Feed Cattle Overseas," CNN, November 5, 2022. https://www.cnn.com/2022/11/05/us/arizona-water-foreign-owned-farms-climate/index.html.
47. Bickel et al., "Contribution of On-Farm Agriculture and Agribusiness."
48. Ibid.
49. Birgit Müller, Leigh Johnson, and David Kreuer, "Maladaptive Outcomes of Climate Insurance in Agriculture," *Global Environmental Change* 46 (2017): 23–33. doi:10.1016/j.gloenvcha.2017.06.010.
50. "EWG's Farm Subsidy Database," farm.ewg.org. https://farm.ewg.org/region.php?fips=04021.
51. Ruixue Wang, Roderick M. Rejesus, and Serkan Aglasan, "Warming Temperatures, Yield Risk and Crop Insurance Participation," *European Review of Agricultural Economics* 48, no. 5, December 2021: 1109–31, https://doi.org/10.1093/erae/jbab034.

52. Stephen Robert Miller, "What Should Farmers Grow in the Desert?," *Mother Jones*, February 13, 2022. https://www.motherjones.com/environment/2022/02/colorado-river-desert-guayule-latex-arizona/.
53. David Yaffe-Bellany, and Michael Corkery, "Dumped Milk, Smashed Eggs, Plowed Vegetables: Food Waste of the Pandemic," *The New York Times*, April 11, 2020, sec. Business. https://www.nytimes.com/2020/04/11/business/coronavirus-destroying-food.html.
54. Interview with Leslie Meyers, manager for the Phoenix Area Office of the U.S. Department of Interior's Bureau of Reclamation, July 24, 2019.
55. "Jails & Prisons—Eloy, AZ (Inmate Rosters & Records)," www.countyoffice.org. https://www.countyoffice.org/eloy-az-jails-prisons/.
56. Matt Weisner, "Reckoning Ahead for Arizona as Water Imbalance Grows on Colorado River," *Water*, May 18, 2017. https://deeply.thenewhumanitarian.org/water/articles/2017/05/18/reckoning-ahead-for-arizona-as-water-imbalance-grows-on-colorado-river. Matthew S. Lachniet, Rhawn F. Denniston, Yemane Asmerom, and Victor J. Polyak, "Orbital Control of Western North America Atmospheric Circulation and Climate over Two Glacial Cycles," *Nature Communications* 5, no. 1 (May 2, 2014). https://doi.org/10.1038/ncomms4805.
57. Matthew S. Lachniet, Rhawn F. Denniston, Yemane Asmerom, and Victor J. Polyak, "Orbital Control of Western North America Atmospheric Circulation and Climate over Two Glacial Cycles," *Nature Communications* 5, no. 1 (2014). https://doi.org/10.1038/ncomms4805.
58. Guadalupe Sanchez, and John Carpenter, "Pleistocene and Holocene Adaptations in the Sonoran Desert," 2016. https://www.academia.edu/21411278/Pleistocene_and_Holocene_Adaptations_in_the_Sonoran_Desert_2016_.
59. "Phoenix Data Center Market," www.datacenters.com. https://www.datacenters.com/locations/phoenix#:~:text=There%20are%20currently%2021%20providers.
60. "Arizona State Sales Tax Incentives at H5 Data Centers Phoenix," h5datacenters.com. https://h5datacenters.com/h5-data-phoenix-tpt.html.

61. "Semiconductor Companies in Arizona," Vintage Computer Chip Collectibles, Memorabilia & Jewelry, n.d. https://www.chipsetc.com/semiconductor-companies-in-arizona.html.
62. Roger Cremades, Anabel Sanchez-Plaza, Richard J. Hewitt, Hermine Mitter, Jacopo A. Baggio, Marta Olazabal, Annelies Broekman, Bernadette Kropf, and Nicu Constantin Tudose, "Guiding Cities under Increased Droughts: The Limits to Sustainable Urban Futures," *Ecological Economics* 189 (November 2021): 107140. https://doi.org/10.1016/j.ecolecon.2021.107140.
63. "Overview of Arizona's Climate Change Preparations," Georgetown Climate Center, 2017. https://www.georgetownclimate.org/adaptation/state-information/arizona/overview.html#state-agency-plans.
64. Md Abu Bakar Siddik, Arman Shehabi, and Landon Marston, "The Environmental Footprint of Data Centers in the United States," *Environmental Research Letters* 16, no. 6 (May 21, 2021): 064017. https://doi.org/10.1088/1748-9326/abfba1.
65. Tom Scanlon, "Mesa Sees Water Cost Increases amid Drought," *East Valley Tribune*, September 22, 2021. https://www.eastvalleytribune.com/news/mesa-sees-water-cost-increases-amid-drought/article_82999310-1ac4-11ec-adf7-ebb6ac90ae66.html.
66. Siddik et al., "The Environmental Footprint of Data Centers in the United States."
67. Daniel Burillo, "Electric Power Infrastructure Vulnerabilities to Heat Waves from Climate Change," Arizona State University, August 2018. https://keep.lib.asu.edu/_flysystem/fedora/c7/201631/Burillo_asu_0010E_18222.pdf.
68. "Arizona Profile," US Energy Information Administration, n.d. https://www.eia.gov/state/print.php?sid=AZ.
69. "During a Global Pandemic, Environmental Inequality Lingers in Arizona's Climate Crisis," *Latino USA*, May 11, 2020. https://www.latinousa.org/2020/05/11/arizonaclimate/.
70. Robin, Cooper, "The Impacts of Extreme Heat on Mental Health," *Psychiatric Times*, July 29, 2019. https://www.psychiatrictimes.com/view/impacts-extreme-heat-mental-health.

71. D. Burillo, "Electric Power Infrastructure Vulnerabilities to Heat Waves from Climate Change," Ph.D. thesis, Arizona State University, Arizona State University, 2018.
72. "Phoenix Endures 145 Days of 100-Degree Heat, Breaking Long-Standing Record," AccuWeather, October 14, 2022. https://www.accuweather.com/en/weather-forecasts/phoenix-sets-new-record-for-most-100-degree-days/831767.
73. Katherine R. Clifford, "Natural Exceptions or Exceptional Natures? Regulatory Science and the Production of Rarity," *Annals of the American Association of Geographers* 112, no. 8 (June 7, 2022): 2287–304. https://doi.org/10.1080/24694452.2022.2054768.
74. "Industry Market Research, Reports, and Statistics," www.ibisworld.com, n.d.. https://www.ibisworld.com/united-states/economic-profiles/arizona/#:~:text=In%202022%2C%20Arizona.
75. E. Gregory McPherson and Renee A. Haip, "Emerging Desert Landscape in Tucson," *Geographical Review* 79, no. 4 (October 1989): 435. https://doi.org/10.2307/215117.
76. Wood, Patel & Associates, Inc., "Master Water Report for Development Unit 2 at Eastmark," June 30, 2020. https://www.mesaaz.gov/home/showpublisheddocument/43386/637587478365870000.
77. Kathleen Ferris and Sarah Porter, "The Myth of Safe-Yield: Pursuing the Goal of Safe Yield Isn't Saving Our Groundwater," Arizona State University: Kyle Center for Water Policy, May 2021. https://morrisoninstitute.asu.edu/sites/default/files/the_myth_of_safe-yield_0.pdf.
78. Michael Kiefer, "Babbitt's Secret Growth-Control Plan," *Phoenix New Times*, March 2, 2000. https://www.phoenixnewtimes.com/news/babbitts-secret-growth-control-plan-6417812.
79. "Arizona Legislature Wants Feasibility Study for Long-Distance Pipeline to Replenish Colorado River Supply," mohavedailynews.com. May 11, 2021. https://mohavedailynews.com/news/131764/arizona-legislature-wants-feasibility-study-for-long-distance-pipeline-to-replenish-colorado-river-supply/.

80. Tony Davis, "Once Again, Arizona Hopes to Import Out-of-State Water in Face of Crisis," *Arizona Daily Star*, May 29, 2021. https://tucson.com/news/local/once-again-arizona-hopes-to-import-out-of-state-water-in-face-of-crisis/article_c47bf80a-be39-11eb-918b-13b88dd52f2f.html#tracking-source=article-related-bottom.
81. Jennifer E. Zuniga, "The Central Arizona Project," Bureau of Reclamation, 2000. https://www.usbr.gov/projects/pdf.php?id=94.
82. "New AZ Law Puts $1.2B toward Finding New Water Sources, Conserving," *Arizona Daily Star*, July 6, 2022. https://tucson.com/news/local/subscriber/new-az-law-puts-1-2b-toward-finding-new-water-sources-conserving/article_3762ae70-fd47-11ec-a0d6-2f60ae6d0747.html#tracking-source=article-related-bottom.
83. Courtney Holmes, "Transparency Concerns Surround Arizona Desalination Deal," ABC15 Arizona in Phoenix (KNXV), December 21, 2022. https://www.abc15.com/news/state/transparency-concerns-surround-arizona-desalination-deal.
84. Jamie McEvoy and Margaret Wilder, "Discourse and Desalination: Potential Impacts of Proposed Climate Change Adaptation Interventions in the Arizona–Sonora Border Region," *Global Environmental Change* 22, no. 2 (May 2012): 353–63. https://doi.org/10.1016/j.gloenvcha.2011.11.001.
85. Edward Lohman, "Yuma Desalting Plant: 2003," *Southwest Hydrology*, May 2003. https://pdfsecret.com/download/the-yuma-desalting-plant-southwest-hydrology_59ffaa49d64ab28ae26ba50f_pdf.
86. D. Rose, "Hohokam," www.arizonaruins.com, February 2014. http://www.arizonaruins.com/articles/hohokam/hohokam.html.

About the Author

Stephen Robert Miller is an award-winning journalist whose work has appeared in *National Geographic*, *The Washington Post*, *The Guardian*, *Discover Magazine*, *Audubon*, *Sierra*, and many others. A former senior editor at *YES! Magazine*, he studied journalism at the University of Arizona and was a Ted Scripps Fellow at the University of Colorado's Center for Environmental Journalism. He lives with his wife and son in Colorado. This is his first book.

Index

Page numbers followed by "f" indicate images.